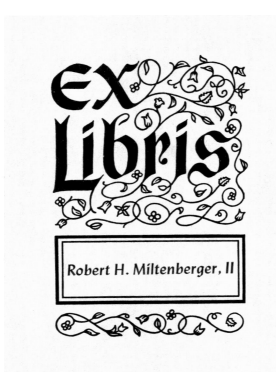

Carbines of the U.S. Cavalry 1861-1905

by
John D. McAulay

ANDREW MOWBRAY PUBLISHERS • P.O. Box 460 • Lincoln, Rhode Island 02865 USA

LIBRARY OF CONGRESS
CATALOG CARD NO.: 95-080556
    John D. McAulay
        Carbines of the United States Cavalry 1861-1905
    Lincoln, RI: ANDREW MOWBRAY INCORPORATED—*PUBLISHERS*
    pp. 144

ISBN: 0-917218-70-1

©1996 by John D. McAulay

All rights reserved. No part of this book may be reproduced in any form or by any means without permission in writing from the author and the publisher.

To order more copies of this book, call 1-800-999-4697.

Printed in the United States of America

This book was designed and set in type by Jo-Ann Langlois. The typeface chosen was Times Roman.

1 2 3 4 5 6 7 8 9 10

This book is dedicated
to the officers and men
of the U.S. Cavalry
by whom these arms were carried
in protection of their country.

ACKNOWLEDGMENTS

I am indebted to the following people who helped make this book a reality.

My research pals Paul Davies and Jerry Coates were of great help. Jerry for his information on the 1862 Kentucky cavalry issues and Paul, for his discovery of the 1861 policy to issue only ten carbines per company per cavalry regiment. My good friend Herb Peck, Jr. provided photos from his collection, as he has for my previous works. I am much in his debt. Larry Jones, Jr. supplied the photograph that was taken at Fort Sill, about 1871, of a 10th Cavalryman armed with an altered Sharps carbine. Randy Hackenburg and Michael Winey at the Army War College, Carlisle Barracks were of tremendous assistance on many of the photos from the Civil War period to the Krag carbines photos in the Philippines. Harry Hunter, Curator of the Division of American History, Museum of American History, Smithsonian Institution and Robert Fisch, Museum Curator, U.S. Military Academy helped acquire photos from those two great collections. Illustrations were also provided by the Custer Battlefield, Springfield Armory, National Archives and the Arizona Historical Society. Additional thanks go to my friend Hubert Lum for photos of the 1855 Joslyn and 1st Model Merrill.

Thanks to Anella Oliva, Mike Niedringhaus and Don Bloomer who were most patient in helping the author learn the ropes of the computer used to type the Trapdoor and Krag sections of the manuscript. Finally, thanks to Jean Jackson who typed the majority of the manuscript.

INTRODUCTION

This is the story of the carbine in combat service with the United States Cavalry from the outset of the Civil War, in April 1861, through the introduction into the field of the '03 Springfield Rifle in 1905. During this 45-year period, over 60 different types and varieties of carbines were available for field service — the bulk of this number being issued during the Civil War.

At the outbreak of the Civil War, the cavalry consisted of five regiments of regulars and a sixth that was formed during the war's first year. These six regiments, plus a large number of volunteer cavalry and mounted infantry regiments formed during the conflict, were to make up an important part of the Union effort. Because of the great shortage of arms at the outbreak of fighting, a policy of only ten carbines per company per regiment was adopted by the Ordnance Department in December of 1861. This policy contributed generously to the oft-discussed ineffectiveness of the Union cavalry during the first two years of the war.

The Sharps carbine was the mainstay of the Union cavalry. It wasn't until the spring of 1865 that the Spencer carbine was issued in quantities nearly equal to the Sharps. On the first day at Gettysburg, July 1, 1863, the bulk of Buford's cavalry was armed with the Sharps carbine. By 1865, the Spencer was to play a dominant part in the outcome of the spring cavalry campaign. At Selma, Alabama, on April 2, 1,550 dismounted Union cavalrymen attacked a fortified Confederate position. The attack proceeded across 600 yards of open ground. When, within a short distance of the main Confederate breastworks, the cavalrymen opened up with their seven-shot, repeating Spencers, they easily carried the Southern position.

While the Spencer and Sharps were the envy of the cavalry, other carbines such as the Gallager, Hall and Starr were to be given the "thumbs down" as totally worthless. The Hall was rated by an officer of the 2nd Missouri Cavalry (Merrill's Horse) as acceptable compared to squirrel guns and shotguns but totally worthless against the better breechloaders.

During the post-war period, the Spencer and the Sharps (first the percussion and then those altered to centerfire) were the major issues to the cavalry until the adoption of the .45-70 Trapdoor Springfield in 1873. For the next twenty years, the Trapdoor was the standard arm of the United States Cavalry. It was during this span of time that the last great Indian Wars took place — the most famous battle being, of course, the Custer debacle at the Little Bighorn. Here, Custer and nearly half of the 7th Cavalry were wiped out by the Sioux and their allies on June 25-26, 1876. During the 1880s, in campaigns against the Apache leader Geronimo, the cavalry was issued a few hundred magazine-fed, five-shot Hotchkiss carbines in addition to the Trapdoors. Most of the Hotchkiss carbines were in the hands of the 4th Cavalry.

Finally, in 1896, the cavalry was rearmed with the five-shot, magazine-fed .30-40 Krag. It was the Krag that would see action first in Cuba, later in the Philippines and then in the Boxer Rebellion. The replacement of the Krag with the '03 Springfield seems a convenient and logical place to conclude my book. The '03, while developed as a rifle, was a general issue weapon; it eliminated the need for a separate and distinct cavalry shoulder arm. The 20th century was well underway, and the significance of horse cavalry in military affairs was already becoming diminished. The era of the cavalry carbine was at an end.

John D. McAulay

TABLE OF CONTENTS

I. *The Civil War Period* ◆ *1861-1865*

A. The Early Years, 1861-1862
Carbines in the Regular Cavalry at the beginning of the war...........11
Inventory at Washington Arsenal December 18, 1861...........11
Policy of December 18, 1861 to issue only 10 carbines per company...........12
1861 State of Rhode Island procurement of Burnside Carbines...........12
1861 State of Ohio procurement of Sharps Carbines...........12
December 28, 1861 skirmish between N.B. Forrest Confederate Cavalry and the 3rd
    Kentucky Cavalry...........12
1st and 2nd California Cavalry field service with Sharps Carbine 1862...........13
1862 field service in Virginia and Kentucky...........12
1862 State of Kentucky issuance by regiment...........14
1862 State of Ohio issuance by regiment...........14
1862 field report of unsatisfactory performance of Gallager Carbines...........14
Colonel Taylor, 5th Ohio, requesting his regiment be issued Wesson Carbines...........17
December 1862 Report of carbines in the various cavalry regiments...........17-21

B. The Turning Point, 1863
Lindner Carbines for 8th West Virginia Mounted Infantry...........23
6,000 Amoskeag Lindner Carbines Court of Claim Case...........23
1863 Field Reports...........24
    Hall...........24
    Model 1862 Joslyn...........25
    Burnside...........25
    Gibbs...........26
    Merrill...........26
    Sharps & Hankins...........26
    Gallager...........27
    Smith...........29
    Spencer...........29
1863 Cavalry Battles
    Grierson's Raid...........29
    Brandy Station...........31
    Gettysburg...........33
    Morgan's Raid...........34
    Chickamauga...........35
September 1863 Report of carbines in the various regiments...........35

C. The Year of Attrition, 1864
Carbines in Arsenal & Depot Storage January 2 and November 5, 1864...........39
1864 Field Reports
    Spencer...........39
    Sharps...........40
    Ballard...........41
    Burnside...........42
    Starr...........42

   Gallager 43
   Union (Cosmopolitan or Gwyn & Campbell) 44
  1864 Campaigns and Battles
   Yellow Tavern 45
   June 30 Army of the Potomac Cavalry Corps Ordnance Stores 46
   Atlanta Campaign 48
   Valley Campaign 48
   Adobe Walls 50
  1864 Confederate Inspection Reports for Bedford Forrest 51
  Summer & Fall inventory reports of various cavalry and mounted infantry regiments 52

D. The End of Hostilities, 1865
  Eastern & Western Military Dept. March 1, 1865 Carbine Inventory 57
  Appomattox Campaign (Dinwiddie Court House, Five Forks, Saylor's Creek) 58
  Wilson's Selma Campaign 60
  Regimental Inventories of Sharps, Burnside, Spencer Carbines 63
  Blakeslee Cartridge Boxes 65-66
  Field Service (Spencer, Ballard, Rimfire Gallager, 2nd Model Maynard, E.G. Lamson's Ball &
   Palmer, Remington, Rimfire Starr, Triplett & Scott, Warner) 65
  Spencers for 6th U.S. Cavalry 74

E. Government Disposals
  Sales at Harpers Ferry - September 1865 75
  Sales at Leavenworth Arsenal - November 1867 75
  Sales of Starr Carbines - 1865-1869 76
  Sales to Argentina - January 1869 76
  Ordnance Sales at St. Louis Arsenal - April 1869 77
  Sale of Spencer Carbines, New York Agency - October 1870 77
  Disposal of Spencer Ammunition from New York Agency - September-December 1870 78
  State of Kentucky, Small Arms Received at New York Agency - October 1871 79
  Percussion Carbines Sales - 1871-1876 79
  Sale of Ballard Carbines - October 1882 80
  Sales from New York Agency - June 1901 80
  Carbine Ammunition Sales - 1901 81

II. *The Indian Wars* ◆ *1866-1891*

A. Post-War Period, 1866-1873
  June 30, 1866 Inventory in U.S. Cavalry 85
  Carbines in Arsenal Storage, June 1866 86-87
  Issues to the new Cavalry Regiments 88
  2nd U.S. Cavalry issues of Spencer Carbines, September 27, 1866 88
  Fetterman Massacre 88
  Cavalry Field Service against the Indians, 1867 90
  June 30, 1867 inventory in U.S. Cavalry 92
  Serviceable Spencer Carbines in storage, January 1869 92
  Alteration of Percussion Sharps to centerfire 93
  Serviceable carbine ammunition in storage, October 1870 94
  December 1870 inventory in U.S. Cavalry 95
  Spencers and Sharps in field service 1871-1873 95
  Experimental Carbine M1870 in Field Service, 1872-1873 97-99

B. Trapdoor Era, 1874-1895

- 45-70 Springfield Carbines in Field Service, March 1874 ... 101
- 45-70 Springfield Carbines in Field Service, September 1874 ... 101
- First Engagements with the Springfield 45-70 ... 107
- Little Bighorn, June 25-26, 1876 ... 102
- Ordnance Stores issued to the 7th Cavalry 1876-1877 ... 105
- 1st & 2nd Model Hotchkiss Carbines in Field Service ... 105-106
- 2nd Model Hotchkiss Carbines issued to the 4th Cavalry 1881 ... 106
- Model 1877 Springfield Carbines issued between 1881-1886 ... 108
- Geronimo Apache Campaign of 1885-1886 ... 108
- Model 1886 Experimental Carbine ... 110
- Ghost Dance, 1890-1891 ... 112
- Troop Issues of Springfield Carbines, 1890-1893 ... 115

III. *Foreign Conflicts* ◆ *1898-1905*

A. U.S Magazine Krag, 1892-1905

- Adoption of the Krag, September 15, 1892 ... 119
- Carbines on hand at Springfield Armory, March 31, 1898 ... 120
- Formation of the Rough Riders, May 1898 ... 120
- Issues of Ordnance Stores from Tampa, Florida Ordnance Depot ... 123
- Cavalry battle at Las Guasimas, Cuba, June 24, 1898 ... 123
- Cavalry battle at San Juan Ridge, Cuba, July 1, 1898 ... 124
- Ordnance Stores issued during the Spanish-American War ... 126
- Model 1898 Krag Carbines sent to the Philippines, 1899 ... 127
- Krag Carbines in Storage, Springfield Armory, September 30, 1899 ... 127
- Krag Carbines issued 1902 ... 129
- Boxer Rebellion, 1900 ... 129

Appendix A — Union Regimental Summary 1861-1866 ... 132

Appendix B — Confederate Regimental Summary ... 134

Bibliography ... 136

Index ... 139

The Civil War Period
1861-1865

On April 22, 1861, the State of Rhode Island purchased 80 Second Model Burnside carbines from the Bristol Firearms Co. These carbines were first issued to the 1st R.I. Infantry. The two individuals shown here are from the 1st. These arms were later issued to the Rhode Island cavalry.

The Herb Peck Jr. collection

THE EARLY YEARS
◆ 1861-1862 ◆

As the year 1860 came to a close, the nation was quickly drifting to open civil war. South Carolina had already seceded and other Southern States were proceeding on a similar course of action.

The regular U.S. Army was listing an authorized strength of 18,093 with an actual strength on hand of 16,367 officers and enlisted personnel as of December 31, 1860. In the five regiments of mounted troops, four of its colonels and three lieutenant colonels resigned their commissions and tendered their services to the Confederacy.[1]

At the outbreak of the conflict in April 1861, the quantity of inventory on hand for the cavalry consisted of only 4,076 carbines; 27,192 revolvers; and 16,933 sabres.[2] At this time, the troops were armed as follows:

1st Dragoons	M1853 Sharps Carbines
2nd Dragoons	M1853 and NM1859 Sharps
Mounted Rifles	M1841 Rifle and 1st M. Maynard
1st Cavalry	NM 1859 Sharps and 1st M. Maynard
2nd Cavalry	NM 1859 Sharps Carbines

On May 4, Congress authorized ten new regiments of which one was cavalry; three months later on August 3, by an Act of Congress, six regiments were redesignated as cavalry with the following name changes:

Prior to Change	After Redesignation[3]
1st Dragoons	1st U.S. Cavalry
2nd Dragoons	2nd U.S. Cavalry
Mounted Rifles	3rd U.S. Cavalry
1st U.S. Cavalry	4th U.S. Cavalry
2nd U.S. Cavalry	5th U.S. Cavalry
3rd U.S. Cavalry	6th U.S. Cavalry

With the defeat at Bull Run in July, it became apparent that this was not going to be a short war. Up to this time, the government had been reluctant to accept volunteer cavalry regiments into federal service. Part of the cause for not mustering in the volunteer cavalry was the price tag of $500,000 to equip and man a regiment of 1,200 men. However, by the first of September, 31 cavalry regiments had been mustered into federal service and, by year's end, this total had risen to 82 regiments and 90,000 mounted troops.[4]

Arming this large cavalry force was a major concern for the Ordnance Department. Without a sufficient quantity of breechloaders for issue, units such as the 4th Iowa and 7th Pennsylvania Cavalry had to settle for second-class rifles. Captain Eugene A. Carr, formerly Captain of the 1st U.S. Cavalry and now, in September, Colonel of the 3rd Illinois, requested the 1st Model Maynard Carbine for his regiment but had to settle for Halls. The shortage of cavalry arms was so acute that on the 18th of December the Washington Arsenal had only the following cavalry ordnance on hand[5]:

983 - Cavalry Sabres	2 - Burnside
3,241 - French Pistols	1 - Joslyn
227 - .36 cal. Starr Pistols	92 - Gallager
150 - Allen Pistols	6 - Symmes
723 - Savage Pistols	6 - Schroeder
100 - Whitney Pistols	157 - Sharps
100 - Beal Remington Pistols	36 - Smoothbore Musketoons
84 - Colt Dragoon Pistols	

Because of a lack of carbines for the cavalry, George Stoneman, Chief of Cavalry, directed that only ten carbines were to be issued per company. This policy was set forth in a letter to General Ripley dated December 18:

Headquarters, Office of Chief of Cavalry
Washington, D.C., Dec. 18, 1861

General Ripley
Chief of Ordnance

Sir:
The volunteer cavalry thus far with one or two exceptions have been armed with pistols and sabres and ten carbines to each company.

This method of arming has been adopted because of the scarcity of carbines and because it was thought that for volunteer cavalry it would not be wise to put too many arms in the hands of inexperienced men.

I have made it a rule to withhold pistols from regiments which are armed with carbines.

Very Respectfully Your Obt. Servant
George Stoneman
Brigadier General
Chief of Cavalry[6]

Stoneman's policy was to effect such cavalry regiments as the 1st Vermont, 5th Ohio and 1st, 3rd and 6th Pennsylvania.

The states tried to help alleviate these shortages by purchasing whatever was available on the open market. The quantities to be found were small. On April 22, the State of Rhode Island received the following items from the Bristol Firearms Co:[7]

80 Burnside Carbines at $35 [ea.] $2,800.00
4,010 cartridges at $21/1,000 . . . . 160.40
250 Percussion Caps 25
Total **$2,960.65**

These 80 2nd Model Burnsides were issued to the 1st Rhode Island Infantry and used by the regiment at the Battle of Manassas. After the regiment was mustered out in August, the Burnsides were issued to the Rhode Island Cavalry.

Also in April, the State of New York purchased 350 Sharps carbines from the firm of Schuyler, Hartley and Graham while the State of Ohio ordered 1,000 New Model 1859 Sharps carbines directly from the Sharps Rifle Mfg. Co. It is interesting to note that the State of New York turned around and sold 347 of their 350 Sharps carbines to the federal government.

When the 1st Dragoons departed California for the war back East, they turned in their Model 1853 Slantbreech Sharps. These carbines were then issued to the 1st and 2nd California Cavalry in late 1861. The turned-in arms from the 2nd Dragoons and 1st U.S. Cavalry at Leavenworth, Kansas were made available for the volunteers. Captain Reno, Commanding Officer at Leavenworth Arsenal on September 24, issued 23 Hall, 20 Greene and 7 Maynard carbines to the 2nd Kansas Infantry. The 2nd had been formed to help repel a Confederate advance into the state. The regiment was on duty for a month and mustered out at the end of October. The carbines issued to them were returned to the arsenal on November 1.[8]

Throughout the later part of the year, the Union and Confederate cavalries met in several small skirmishes. One of these fights occurred on December 28 between Confederate cavalry led by Colonel Nathan B. Forrest and the 3rd Kentucky (Union) under the command of Major Eli H. Murray. Major Murray's men were armed with pistols and sabres while Forrest's cavalry had Sharps and Maynard carbines. In his report, Forrest states that he used a Maynard carbine to fire at the rear guard of the Union cavalry and also dismounted part of his command to act as skirmishers with their carbines. In the action, the 3rd Kentucky suffered eight killed and 13 captured out of 168 engaged. Forrest's losses were given as two killed and three wounded out of 300.[9]

1862

With the Regulars back East, it was the Western volunteer cavalry regiments which backfilled the void left by the Regulars' departure. The volunteers' first task was to repel the Confederate excursion into Arizona and New Mexico and later to protect the settlers from hostile Indian raids.

A portion of Carlton's California Column, which marched from Arizona to Northwest Texas between April 8 and September 20, were elements of the 1st and 2nd California Volunteer Cavalry. While the Column saw little Confederate activity, the Indians were much more of a problem. Late in the day on July 15, Sergeant Mitchell and five privates from Co. B, 2nd California Cavalry were attacked by Apaches near Apache Pass Station. In the attack,

The Early Years, 1861-1862

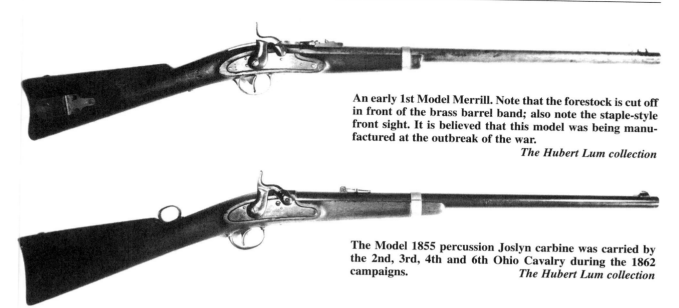

An early 1st Model Merrill. Note that the forestock is cut off in front of the brass barrel band; also note the staple-style front sight. It is believed that this model was being manufactured at the outbreak of the war.
The Hubert Lum collection

The Model 1855 percussion Joslyn carbine was carried by the 2nd, 3rd, 4th and 6th Ohio Cavalry during the 1862 campaigns.
The Hubert Lum collection

Private John W. Teal became separated from the others and was attacked by fifteen Apaches. When his horse was killed, Teal lay flat on the ground and held off the hostiles with his slantbreech Sharps and a Colt revolver. He was able to reach camp unharmed. In the attack, one cavalryman had been wounded. In late July, when the Column left Tucson for the Rio Grande, General Order No. 10 called for the wagon train to handle forty day's rations for the command, plus 40,000 rounds of rifle musket ammunition, 30,000 rounds for the Sharps carbines and 20,000 rounds for the navy-size Colt Revolvers.[10]

Back East, Stoneman's policy of only ten carbines per company was having a negative effect on the cavalry's ability to fight the Confederate cavalry. Near Milford, Virginia on June 24, Companies B and M of the 1st Michigan Cavalry commanded by Major Charles H. Town, attacked a portion of a Louisiana mounted rifle regiment. In this attack, the 1st had but 20 carbines while the Confederates were armed with long range rifles. Since Major Town could not get close enough to the enemy to use his revolvers, he broke off the action. Town's frustrations with the lack of sufficient carbines can be seen in his official report when he states: "it was next to impossible to encounter our foes successfully as he is never in position where sabre and pistol can be used."[11]

On Jeb Stuart's second day out (June 13) on his ride around McClellan's Army, Stuart attacked Companies C and F of the 5th U.S. Cavalry at Gibson's Mill. The regulars, commanded by Captain William B. Royall, suffered four killed, twelve wounded and several captured. In Lieutenant Edward H. Lieb of Co. F's report of the engagement, he states:

I felt most seriously the superiority of the enemy, who were armed with rifles and shot-guns, and had my command been furnished with carbines I would have been able to do him more injury and hold him longer in check. After I had emptied all of my pistols I drew sabers and endeavored to charge.[12]

Later in the year at Perryville, Kentucky, the Union cavalry consisted of the 2nd Michigan, 9th Pennsylvania and 9th Kentucky. They were equipped with a wide assortment of small arms. The 2nd Michigan was armed with Colt revolving rifles while the 9th Pennsylvania was equipped with 41 Sharps and 13 Maynards; and the balance of the regiment (450 men) with only pistols and sabres.[13] The men from Kentucky, the 9th, were armed with 175 Wesson carbines.[14]

The State of Kentucky mustered into Federal service seven cavalry regiments during the year. To arm these troops, the state contracted with Merwin and Bray of New York City for Ballard carbines and with Ben Kittredge & Co. for Wesson carbines and Smith & Wesson revolvers. On August 30, Kentucky received 320 Cosmopolitan carbines from Gwyn & Campbell plus additional deliveries of 891 on December 15. The Cosmopolitans had been purchased at $27 each.[15]

STATE OF KENTUCKY CAVALRY ISSUES — 1862[16]							
	CARBINES			RIFLE	REVOLVERS		
	Wessons	Ballard	Cosmopolitan	Henry	Colt .36	Colt .44	S&W
6th		480			750		
7th	690						736
8th	469	200					
9th	175						
10th						1,200	
11th		300	320				
12th			800	70			
Total	1,334	980	1,120	70	750	1,200	736

The State of Ohio was also issuing small arms from the state inventories as shown in the following chart:

STATE OF OHIO ISSUES — 1862[17]								
	CARBINES			REVOLVERS				
	Sharps	Joslyn	Gallager	Colt	Joslyn	Starr	Remington	Whitney
1st Ohio Cav.	658			1,090			93	
2nd Ohio Cav.		100				700	500	
3rd Ohio Cav.		100					650	50
4th Ohio Cav.		100						
5th Ohio Cav.	120				675			
6th Ohio Cav.		250		71	324			
7th Ohio Cav.			400				1,000	200
1st Squadron	186						184	
2nd Independent Batt'l			100				300	
Total	964	550	500	1,161	999	700	2,727	250

On May 26, a Colonel Thomas Florence of Washington, D.C. offered 3,143 rifled Jenks carbines at $18 each to the Ordnance Department. These Jenks are the same arms purchased by Eastman from the Navy in August 1861 at the cost of $3 each. Ripley felt the Jenks had too many defects to be effective for cavalry use; therefore, he declined the offer the next day. Jenks carbines altered to the Merrill design had been previously offered at $25 each and they also had been refused.[18]

Reports were being received at the Ordnance Department on the effectiveness of the various small arms in service. One of the first complaints received was on the Gallager carbines. The major objection was the difficulty in withdrawing the metallic cartridge from the chamber after firing. A rammer was needed to force the spent case out. The cartridge wrench, which was attached to the screwdriver and provided with the Gallager carbine, was effective in removing the stuck cartridge, but these combination tools had a tendency to become lost in actual field service. Chief of Ordnance for the

The Early Years, 1861-1862

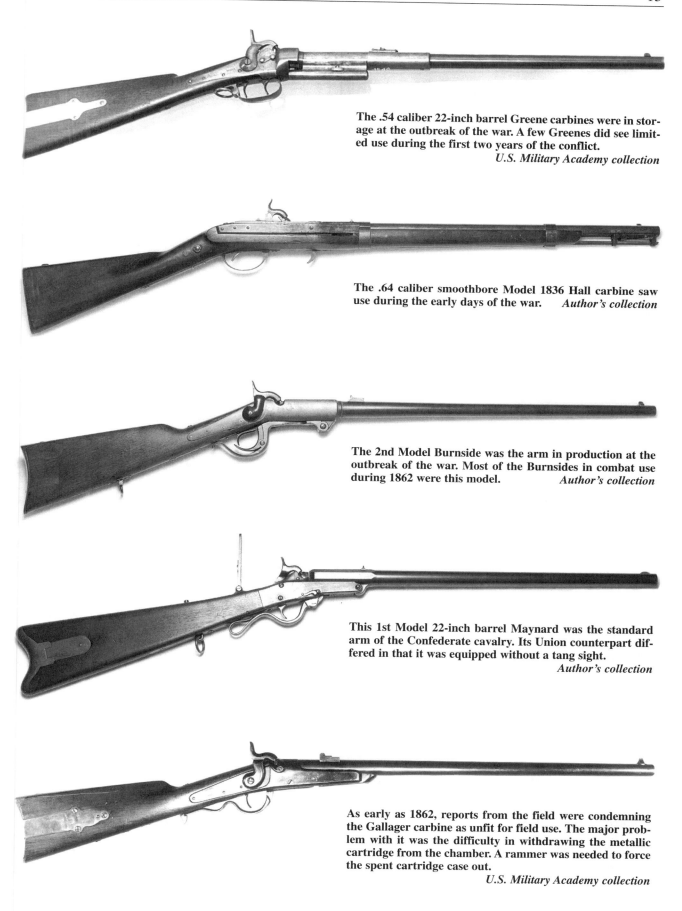

The .54 caliber 22-inch barrel Greene carbines were in storage at the outbreak of the war. A few Greenes did see limited use during the first two years of the conflict.
U.S. Military Academy collection

The .64 caliber smoothbore Model 1836 Hall carbine saw use during the early days of the war. *Author's collection*

The 2nd Model Burnside was the arm in production at the outbreak of the war. Most of the Burnsides in combat use during 1862 were this model. *Author's collection*

This 1st Model 22-inch barrel Maynard was the standard arm of the Confederate cavalry. Its Union counterpart differed in that it was equipped without a tang sight.
Author's collection

As early as 1862, reports from the field were condemning the Gallager carbine as unfit for field use. The major problem with it was the difficulty in withdrawing the metallic cartridge from the chamber. A rammer was needed to force the spent cartridge case out.
U.S. Military Academy collection

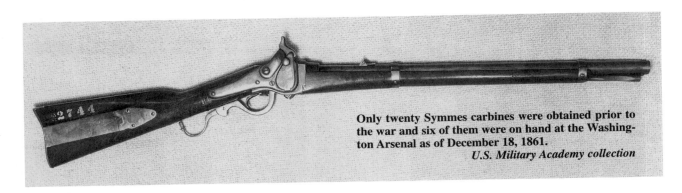

Only twenty Symmes carbines were obtained prior to the war and six of them were on hand at the Washington Arsenal as of December 18, 1861.
U.S. Military Academy collection

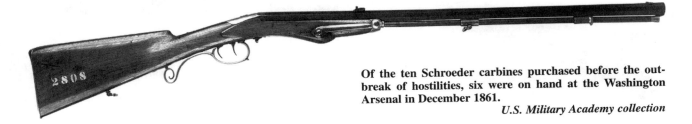

Of the ten Schroeder carbines purchased before the outbreak of hostilities, six were on hand at the Washington Arsenal in December 1861.
U.S. Military Academy collection

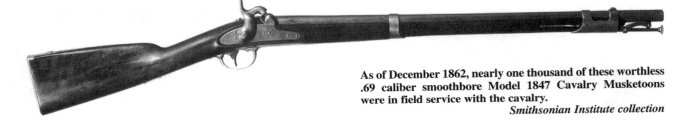

As of December 1862, nearly one thousand of these worthless .69 caliber smoothbore Model 1847 Cavalry Musketoons were in field service with the cavalry.
Smithsonian Institute collection

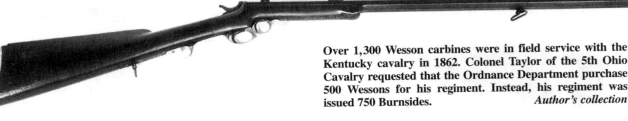

Over 1,300 Wesson carbines were in field service with the Kentucky cavalry in 1862. Colonel Taylor of the 5th Ohio Cavalry requested that the Ordnance Department purchase 500 Wessons for his regiment. Instead, his regiment was issued 750 Burnsides.
Author's collection

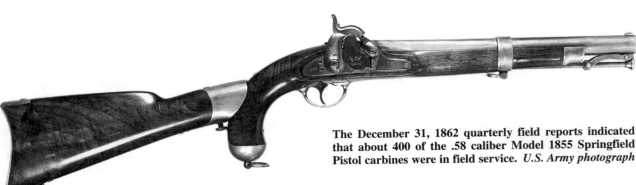

The December 31, 1862 quarterly field reports indicated that about 400 of the .58 caliber Model 1855 Springfield Pistol carbines were in field service. *U.S. Army photograph*

The Early Years, 1861-1862

Department of Ohio, Captain William Hannis states: "I have never met with an officer who had tried the Gallager carbine in the field, without pronouncing it unfit for service."[19]

Many officers tried, on their own, to purchase arms for their regiments. One such officer, Colonel Taylor of the 5th Ohio Cavalry, had found a source (probably Ben Kittredge & Co.) for 500 Wesson carbines at $25 each. In Taylor's colorful letter of October 8th to the Ordnance Department, he requests authority to purchase them. His letter reads:

If the Government will allow me to purchase carbines of Smith & Wesson patent, which I can purchase without any difficulty, at $25 a piece, I will knock the Socks, hats, Boots of any cavalry regiment in the Confederate service, at 500 yards distance, emptying a saddle at every shot.

The difficulty with the Confederate cavalry is that they are all guerrillas, and I never can get near enough to them to charge upon them with sabres. They fire upon us from bushes and then run like the devil, and if I had these carbines, I could kill them at 500 yards at full speed, do me the favor to get permission to purchase 500 of these carbines at the price named and if I do not pay the Government rebel scalps two to one, for the guns given in less then six months, I will forfeit my own scalp. Rebel cavalry are generally armed with Colt revolvers, Colt rifles, Sharps carbines or shot-guns, while the loyal men of the United States Cavalry are armed with sabres and pistols only.[20]

The Ordnance Department felt the price of $25 for the Wesson was too high and, therefore, turned down Taylor's request. In November, Taylor's 5th Ohio was issued 750 Burnside carbines. While these arms were not Wessons, they should have found favor with the colonel.

Regimental Inventories

By year's end, the Ordnance Department had directed that all cavalry, infantry and artillery regiments report their ordnance stores on a quarterly basis starting with the period ending December 31, 1862. While not all regiments reported and others reflected only a partial list of their inventories on hand, these reports do give the reader a good general description of the types and quantities of small arms in field service. The carbines listed in the December report are shown as follows:

Union cavalrymen were not the only ones armed with the Wesson carbine. A Southern cavalryman is shown here with one. *The Herb Peck Jr. collection*

Carbines in the U.S. Cavalry as of December 31, 1862[21]

Sharps	13,645
Maynard	18
Smith	1,922
Wesson	183
Cosmopolitan	492
Ballard	85
Merrill	1,040
Sharps & Hankins	67
Gallager	944
Musketoon	947
Hall	1,909
Pistol Carbine	398
Burnside	895

Continued on page 19

A Union sergeant, probably from Kentucky, armed with a Ballard carbine. *The Herb Peck Jr. collection*

The Early Years, 1861-1862

The following is a breakdown by regiment:

Regiment	Count & Type	Regiment	Count & Type	Regiment	Count & Type
1st U.S.	332 Sharps		80 Hall	2nd Ky.	70 Merrill
	41 Smith	10th Ill.	20 Hall	3rd Ky.	92 Sharps
2nd U.S.	469 Sharps	11th Ill.	19 Sharps	5th Ky.	40 Burnside
	30 Smith		339 Smith		2 Merrill
3rd U.S.	83 Sharps	12th Ill.	57 Burnside	6th Ky.	6 Gallager
	49 Smith	1st Ind.	41 Pistol Carbine	8th Ky.	71 Wesson
	1 Maynard		48 Burnside		85 Ballard
4th U.S.	180 Sharps		121 Hall		98 Gallager
	48 Smith	3rd Ind.	54 Sharps	9th Ky.	113 Wesson
5th U.S.	321 Sharps		58 Gallager	1st La.	294 Sharps
6th U.S.	621 Sharps		80 Burnside	1st Maine	12 Sharps
1st Calif.	355 Sharps	1st Iowa	9 Sharps		116 Burnside
2nd Calif.	601 Sharps		91 Gallager	1st Mass.	102 Sharps
1st Colo.	63 Sharps	2nd Iowa	226 Sharps		285 Smith
2nd Colo.	2 Sharps	3rd Iowa	70 Sharps	2nd Mass.	168 Sharps
1st Conn.	166 Smith	5th Iowa	131 Sharps	1st Md.	75 Sharps
2nd Ill.	338 Sharps		91 Hall	1st Mich.	385 Sharps
	30 Smith	2nd Kan.	46 Hall	2nd Mich.	8 Sharps
3rd Ill.	299 Hall		53 Pistol Carbine	3rd Mich.	37 Sharps
4th Ill.	463 Sharps	5th Kan.	106 Sharps	1st Minn.	115 Sharps
	4 Hall		72 Hall	2nd Mo. State Militia	
5th Ill.	286 Cosmopolitan		2 Maynard		51 Hall
6th Ill.	248 Sharps		27 Musketoon	5th Mo. State Militia	
7th Ill.	399 Sharps		76 Pistol Carbins		51 Musketoon
	252 Smith	6th Kan.	56 Sharps	12th Mo. State Militia	
	206 Cosmopolitan		130 Merrill		138 Musketoon
8th Ill.	440 Sharps		61 Pistol Carbine	13th Mo. State Militia	
9th Ill.	27 Pistol Carbine	7th Kan.	88 Sharps		20 Musketoon

Continued on page 21

Guarding Confederate prisoners in 1862, the soldier on the right is armed with a Sharps carbine.
U.S. Army Military History Institute, MOLLUS

Company G, 1st Massachusetts Cavalry at Edisto Island, South Carolina, in May/June 1862. Three of the troopers are shown with the iron-mounted patchbox Model 1859 Sharps carbine.
U.S. Military History Institute, MOLLUS

Another view of Company G, 1st Massachusetts Cavalry at Edisto Island, South Carolina, in May/June 1862.
U.S. Army Military Institute, MOLLUS

Unit	Qty	Type	Unit	Qty	Type	Unit	Qty	Type
2nd Mo.	299	Hall	4th Ohio	288	Sharps	1st W.V.	5	Pistol Carbine
3rd Mo.	82	Hall	5th Ohio	131	Sharps		36	Sharps
4th Mo.	99	Hall		222	Burnside		198	Smith
	10	Pistol Carbine	6th Ohio	76	Smith	2nd W.V.	66	Pistol Carbine
6th Mo.	227	Hall	1st Penn.	481	Sharps		165	Smith
7th Mo.	304	Hall	2nd Penn.	92	Sharps	2nd Wisc.	6	Sharps
1st N.J.	168	Burnside	3rd Penn.	307	Sharps		25	Hall
1st N.M.	301	Musketoon	4th Penn.	200	Sharps	3rd Wisc.	56	Smith
1st N.Y.	22	Sharps	5th Penn.	464	Gallager		540	Merrill
	105	Gallager	6th Penn.	43	Sharps			
2nd N.Y.	402	Sharps	7th Penn.	108	Smith			
3rd N.Y.	192	Sharps	8th Penn.	295	Sharps			
	48	Burnside	9th Penn.	35	Sharps			
4th N.Y.	19	Merrill		15	Maynard			
5th N.Y.	39	Sharps	11th Penn.	93	Sharps			
6th N.Y.	410	Sharps	13th Penn.	9	Burnside			
7th N.Y.	301	Sharps		122	Gallager			
8th N.Y.	334	Sharps	14th Penn.	224	Burnside			
9th N.Y.	373	Sharps	16th Penn.	200	Sharps			
	45	Smith	1st R.I.	70	Sharps			
	67	Sharps & Hankins	1st Tenn.	146	Sharps			
10th N.Y.	34	Smith		277	Merrill			
11th N.Y.	59	Pistol Carbine		20	Burnside			
1st Ohio	184	Sharps	2nd Tenn.	102	Merrill			
2nd Ohio	1	Sharps	1st Texas	410	Musketoon			
3rd Ohio	53	Sharps	1st Vt.	152	Sharps			
				2	Burnside			

OTHER UNITS

Potomac Home Guard
119 Sharps

McLaughlin Ohio Batt'l
92 Sharps

Sioux City Iowa Cav.
89 Hall

Ringgold Batt'l
100 Sharps

Greenfield Co. Cav. (Penn.)
49 Sharps

Steward Ill. Batt'l
74 Sharps

Footnotes — The Early Years, 1861-62

[1] Stephen Z. Starr, *The Union Cavalry in the Civil War, Vol. 1,* Baton Rouge, LA, 1979, pp. 47 and 58
[2] *Ibid*, p. 66
[3] James A. Sawicki, *Cavalry Regiments of the U.S. Army,* Dumfries, VA, pp. 151-163
[4] Stephen Z. Starr, op.cit., pp. 66 and 78
[5] NARG 156-21
[6] *Ibid.*
[7] NARG 217-759 State Claims Rhode Island
[8] *Ibid*, State Claims Kansas
[9] ORs Series I, Vol. 7, Part I, pp. 62-66
[10] ORs Series I, Vol. 50, Part I, pp. 91 and 131-133
[11] ORs Series I, Vol. 12, Part I, pp. 815-816
[12] ORs Series I, Vol. 11, Part I, pp. 1020-1022
[13] Stephen Z. Starr, *The Union Cavalry in the Civil War, Vol. III,* Baton Rouge, LA, 1985, p. 92
[14] NARG 217-759 State Claims Kentucky
[15] Information on the State of Kentucky purchase of Cosmopolitan carbines courtesy of H. Michael Madaus
[16] NARG 217-759 State Claims Kentucky
[17] 1863 Executive Document State of Ohio 454-455
[18] NARG 156-6
[19] NARG 156-1001
[20] NARG 156-21
[21] NARG 156-110

Joe McCloud, a scout on Sibley's Dakota Expedition in 1863, armed with a Sharps carbine.
The Herb Peck Jr. collection

THE TURNING POINT
◆ 1863 ◆

Nine days into the new year, the New York Arsenal took delivery of 501 Lindner carbines from the Amoskeag Mfg. Co. on an order placed with the company in November of the prior year. In the spring of the year, these carbines were sent to the Wheeling, West Virginia Ordnance Depot for issue. On the 16th and 17th of June, the 8th West Virginia Mounted Infantry was issued Lindner carbines and cavalry sabres from the depot.[1] By the end of the month, 252 Lindners had been issued to the following companies: B-57, D-48, G-45, I-55 and K-47. Three months later, in September, the 8th West Virginia was reflecting the following small arms:

8th West Virginia Mounted Infantry[2]
Small Arms
September 30, 1863

	Lindner Carbines	Enfield Rifles
Co. A	49	
Co. B		58
Co. D	41	
Co. G	42	
Co. H	49	
Co. I	41	
Co. K	36	
In storage	39	
Total	**297**	**58**

The 8th's first action with their Lindners occurred on August 26 and 27 near White Sulpher Springs, West Virginia. In this all-day skirmish with Confederate forces, the 8th suffered two killed, sixteen wounded and three missing.[3] Later in the year, on November 6, in a dismounted attack, the 8th along with the 2nd, 3rd and 10th West Virginia Mounted Infantry regiments were able to drive the Southern defenders off Droop Mountain. In this action, the 8th listed eleven casualties.[4] On January 26, 1864, the regimental designation was changed to the 7th West Virginia Cavalry.

Interested in obtaining a substantial contract, Edward Lindner wrote the Ordnance Department on February 25 requesting an order for 5,000 to 10,000 Lindner carbines at the stated price of $20 each and ammunition at $22 per thousand. In his letter, he states that the Amoskeag Mfg. Co. had the capability of manufacturing 1,000 carbines per month. Nearly two months later, on April 13, an order was given for all the carbines that could be produced in six months, not to exceed 6,000 carbines. Lindner accepted the offer fours days later. Before production could start, Major Hagner, on the 23rd of April, directed the placement of the rear-sight in front of the breech. He also directed that the lockplate be slightly changed in front of the cone so as to have metal to protect the wood. Hagner felt that the Lindners, in current field use, were liable to be split in front of the lockplate with the current design.[5]

It took several months to make the changes directed by Major Hagner. Later in the year, the company was notified that the Lindners in field service were proving to be unsatisfactory. These rumors were later discovered to be unfounded. Finally, on April 5, 1864, E.A. Straw, agent for Amoskeag, submitted a sample carbine to the Ordnance Department. Straw told the ordnance office that the 6,000 Lindner carbines were nearing completion and were ready for inspection. He requested that the ordnance inspectors currently at the factory inspecting rifle-muskets be directed to inspect the Lindner carbines. For some unknown reason, no inspection was ever made even though the ordnance inspectors were at the factory continuously until April 1865.[6]

In May 1865, when the Ordnance Department refused to accept the 6,000 Amoskeag-marked Lindners, the company filed suit for damages. The suit dragged on for many years. In February 1869, the War Department sent the U.S. Attorney General the facts in the case which was under way in the Court of Claims. A year later, on May 7, 1870, depositions were taken at the law offices of Denvor and Pack located at 1115 Pennsylvania Avenue, Washington, D.C.[7] The final trial results are unclear, but it is believed that Amoskeag lost their appeal and that the carbines were disposed of on the European market in the 1870s.

1863 FIELD REPORTS

Hall

The oldest breechloading carbine in the Union cavalry was the Hall. The Hall had been in field service for nearly 30 years. It was generally unpopular with the troops. In his official report on the battle of Helena, Arkansas, dated July 6, Lieutenant Colonel Thomas N. Pace of the 1st Indiana Cavalry stated that "more than half the regiment threw away their carbines, many of them being unserviceable having been condemned by the United States inspecting officer some time since, and supplied themselves with Enfield rifles captured from the enemy."[8] During their dismounted attack on the Confederate position, the 1st Indiana suffered two killed, eight wounded and one missing. In this engagement, the regiment was mainly armed with Hall carbines and a few pistol carbines.

The following two officers had different opinions on the merits of the Hall carbine. In September, Major Spellman, 7th Missouri Vol. Cavalry with 202 Hall carbines, considered them unreliable. On the other hand, in December, Colonel Catherwood, 6th Missouri State Militia Cavalry with 290 Halls, stated that they were very effective carbines for cavalry use. The opinion of an officer from the 2nd Missouri Vol. Cavalry (Merrill's Horse) was probably more typical of the view held by most officers that the Hall carbine was a good arm compared to shotguns and squirrel rifles, but among the better breechloaders, there was no comparison.[9]

Model 1862 Joslyn

The only cavalry regiment known to have been issued the Model 1862 rimfire Joslyn carbine was the 19th New York. The regiment had been converted from the 130th Infantry to the 19th Cavalry on August 11. On the first of September, they were issued the Model 1862 Joslyn carbines.

The first field report on the Joslyn was dated October 4 from the 19th New York commanding officer Colonel Alfred Gibbs. Colonel Gibbs felt that the Joslyn was the most accurate carbine he had ever seen and better than the Sharps. Gibbs also gave it good marks for ease in loading and claimed that it did not foul easily. He raised the objection that the breechblock (piston) was highly tempered and, therefore, easily broken. The front sights also had a tendency to come off. The colonel closed his report by stating that he hoped the carbines would prove effective for cavalry service. Two months later, in a follow-up report, Colonel Gibbs writes that the Spencer cartridges were a trifle too large for the chamber and, therefore, difficult to withdraw after firing. The Joslyn metallic cartridges fit well but, like the Spencer ammunition, they also sometimes lacked fulminate in the cartridge. Otherwise, he felt it gave overall satisfactory results.[10]

The officers of the 19th New York did not share the same opinion of the Joslyn as their colonel. Of the six company commanders responding to the Ordnance Department fourth quarter survey report, only one officer felt it gave good results — that officer was Captain Wells of Company G. Captains Cullerton of Co. B, Godfrey of Co. C, Hake of Co. E, Britton of Co. H and Linch of Co. K all regarded the Joslyn as a poor arm.[11] These officers' complaints were that the pistons and caps broke and that extracting the copper cartridge was often difficult. As late as March of the following year, the 19th was still armed with 583 Joslyns.

Burnside

By the fall of 1863, nearly 8,000 Burnside carbines were in field service. Most of the carbines were of the fourth model. These four reports were typical:[12]

Lieutenant Colonel Stedman of the 6th Ohio Cavalry with 179 Burnsides stated in November that they were vastly inferior to the Sharps and felt they should be condemned.

Lieutenant Colonel Ruggles, 3rd Illinois with 114 Burnsides, stated in November that his men liked their Burnsides. They were light to carry, shot

The Turning Point, 1863

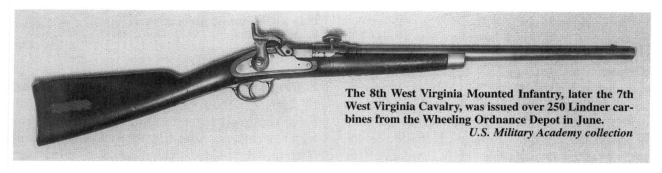

The 8th West Virginia Mounted Infantry, later the 7th West Virginia Cavalry, was issued over 250 Lindner carbines from the Wheeling Ordnance Depot in June. *U.S. Military Academy collection*

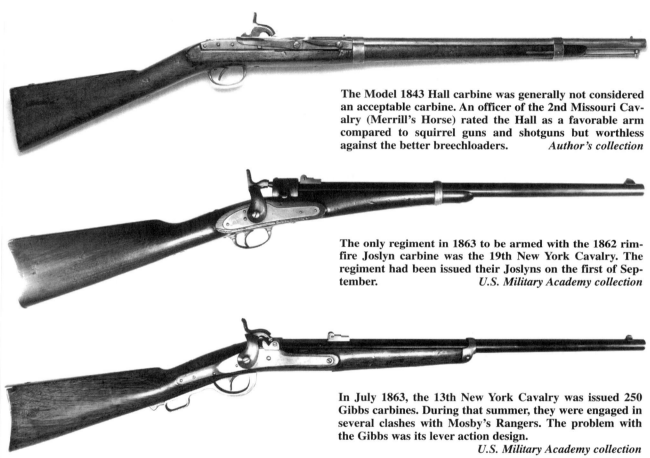

The Model 1843 Hall carbine was generally not considered an acceptable carbine. An officer of the 2nd Missouri Cavalry (Merrill's Horse) rated the Hall as a favorable arm compared to squirrel guns and shotguns but worthless against the better breechloaders. *Author's collection*

The only regiment in 1863 to be armed with the 1862 rimfire Joslyn carbine was the 19th New York Cavalry. The regiment had been issued their Joslyns on the first of September. *U.S. Military Academy collection*

In July 1863, the 13th New York Cavalry was issued 250 Gibbs carbines. During that summer, they were engaged in several clashes with Mosby's Rangers. The problem with the Gibbs was its lever action design. *U.S. Military Academy collection*

well and the cartridges were well protected.

In September, Major Bradley, 7th Kentucky with 215 Burnsides, felt that they were always getting out of order.

Colonel Sipes, 7th Tennessee with 136 carbines, reported in November that he believed the material used to manufacture the carbine barrels was of an inferior quality.

Gibbs

In 1863, three cavalry regiments were armed with Gibbs carbines. They were the 10th Missouri plus the 13th and 16th New York. The first two regiments filed reports from the field and gave the thumbs down on the Gibbs.

The 10th Missouri Vol. Cavalry located at Natchez, Mississippi, in December, was armed with 496 Gibbs carbines. Major Benton, of the 10th, stated that the Gibbs did not meet the requirements of the cavalry. He found that they were often rendered useless after being discharged. Even with the dislike for the Gibbs, it remained in the regimental inventory through most of 1864.

In July, the 13th New York Cavalry, commanded by Lieutenant Colonel Gansevoort, was issued 250 Gibbs carbines in their campaign against Mosby's Rangers. After four months in field service, Colonel Gansevoort requested that the Gibbs no longer be issued to his regiment. The major objection raised by the colonel was "caused by a poor lever action. Its barrel is moved forward and backward in the stock by a lever which passes vertically through the stock. When at all rusted, this lever is difficult to reach so as to *entirely* clean it and, when worked by the guard in that state, is apt to break off in the stock near the barrel."[13] Other problems were the inferior workmanship and material used in its construction plus several of the carbines burst about six to eight inches from the muzzle due to the thinness of the barrels. In the following year, the 13th turned in their Gibbs for Sharps carbines.

Merrill

During the fourth quarter of the year, over 20 officers from nine regiments responded on the Merrill carbines. Their views ranged from the Merrill being very good, to totally unfit for service use. The major complaints were that the stocks broke most often at the wrist area. Other concerns raised were that the sights came loose and the carbines became fouled after a few firings so that it was impossible to open the breech for reloading. The Secretary of War was notified that many of the carbine barrels were bursting; therefore, he directed that all Merrill barrels be proved at the Washington Arsenal. These arsenal-proved barrels will be found with VP, eagle stamped on them.[14]

THE FOLLOWING CHART SHOWS THE OFFICERS' OPINIONS OF THE MERRILL:[15]

Poor to Worthless		**Good to Excellent**	
Capt. Arthur	Co. B — 2nd Kentucky	Capt. Brown	Co. B — 11th Missouri
Capt. Platt	Co. D — 10th Ohio	Capt. Weber	Co. M — 11th Missouri
Capt. Bass	Co. K — 5th Tennessee	Capt. Filkin	Co. E — 10th Ohio
Capt. Howland	Co. C — 1st Wisconsin	Capt. Witt	Co. G — 3rd Tennessee
Capt. Cornstock	Co. F — 1st Wisconsin	Capt. Casen	Co. L — 5th Tennessee
Capt. Robinson	Co. G — 1st Wisconsin	Capt. Senive	Co. C — 7th Indiana
Capt. Tripp	Co. B — 1st Illinois	Capt. Hubbard	7th Indiana
Capt. Wright	Co. D — 7th Indiana	Capt. Daily	Co. L — 7th Indiana
Capt. Skinner	Co. E — 7th Indiana	Capt. Pall	Co. C — 11th Missouri
Capt. Moons	Co. H — 7th Indiana	Capt. Witson	Co. A — 2nd U.S. Colored
Capt. Elliott	Co. M — 7th Indiana		

Sharps & Hankins

One unit reported on the Sharps & Hankins carbine. This unit — the 9th New York Cavalry commanded by Colonel Sackett — was armed with 123 Sharps & Hankins Model 1861 carbines in September. Colonel Sackett had these comments on their effectiveness:

Could the spring that holds the guard on Sharps & Hankins carbines and the catch that draws back the copper after firing be strengthened that arm would certainly be better than Sharps, Burnsides or Starrs, as the ammunition never dampens or breaks and no capping being required. The rapidity of fire makes it a very fine skirmishing weapon.[16]

Gallager

From Murfreesboro, Tennessee, Lieutenant Horace Porter, ordnance officer for the Department of the Cumberland, wrote Ripley on the problems with the Gallager carbines. As of April, Porter stated that Gallagers were being carried by the 9th Pennsylvania, 10th Ohio and 4th Kentucky Vol. Cavalry regiments. A captain from the 10th Ohio reported that in firing 32 Gallagers one day, ten would not fire; the cartridge cases stuck fast in fourteen and three burst. Lieutenant Porter concludes his report by stating that "the Gallager carbine when well made, in the hands of experienced and careful troops, each carrying a wrench in a convenient place for remov-

The Turning Point, 1863

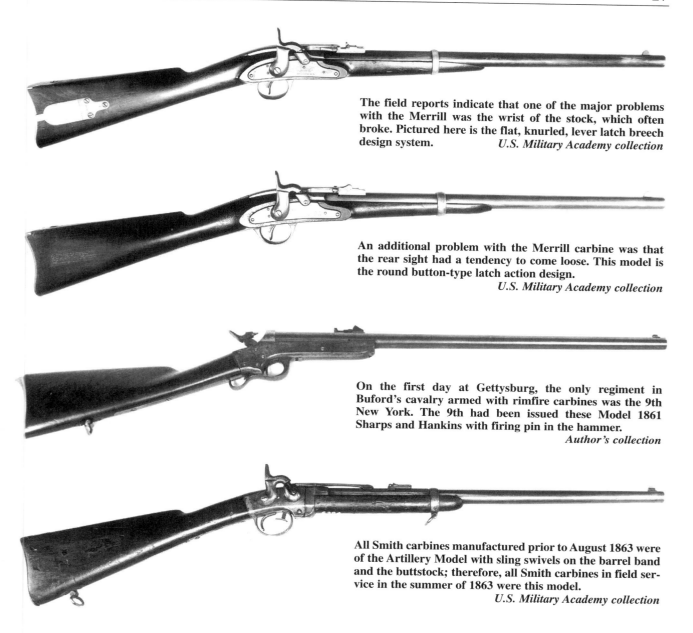

The field reports indicate that one of the major problems with the Merrill was the wrist of the stock, which often broke. Pictured here is the flat, knurled, lever latch breech design system.
U.S. Military Academy collection

An additional problem with the Merrill carbine was that the rear sight had a tendency to come loose. This model is the round button-type latch action design.
U.S. Military Academy collection

On the first day at Gettysburg, the only regiment in Buford's cavalry armed with rimfire carbines was the 9th New York. The 9th had been issued these Model 1861 Sharps and Hankins with firing pin in the hammer.
Author's collection

All Smith carbines manufactured prior to August 1863 were of the Artillery Model with sling swivels on the barrel band and the buttstock; therefore, all Smith carbines in field service in the summer of 1863 were this model.
U.S. Military Academy collection

ing cartridge cases, would be a good cavalry arm. But we have no such troops and should give them such weapons as are most serviceable in their hands. — A complaint scarcely ever heard against Sharps or Burnsides and seldom against Smith carbines, and I speak relatively rather than absolutely in regard to Gallagers. I pronounce it an inferior arm."[17]

Smith

The reports on the Smiths ranged from utterly unreliable (1st Massachusetts Cavalry) to the best (10th New York). The 10th New York rated the Smith as the best because it was easy to clean and more durable than the Sharps. They also gave it high ratings for accuracy and range and its ease in loading on horseback. The 2nd Arkansas was of the opinion, concerning the three types of carbines in the regiment, that the Smith was the best for simplicity, accuracy and range.[18] Its durability was also cited by Lieutenant Colonel Dodds, 1st Alabama (U.S.).

All Smith carbines manufactured prior to August 1863 were of the Artillery Model with the sling swivels on the barrel band and on the bottom of the buttstock. These carbines were manufactured at the Massachusetts Arms Co. All Smith carbines in field service through September of 1863 were of this model.[19]

The 1st Massachusetts Cavalry's dislike for the Smith went back to a skirmish with Confederate

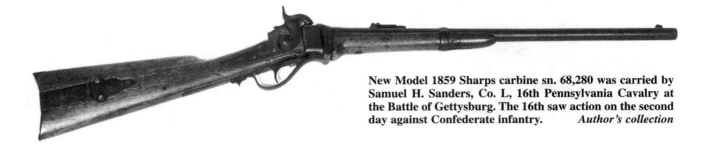

New Model 1859 Sharps carbine sn. 68,280 was carried by Samuel H. Sanders, Co. L, 16th Pennsylvania Cavalry at the Battle of Gettysburg. The 16th saw action on the second day against Confederate infantry. *Author's collection*

cavalry in November of the previous year. The action occurred when they were crossing a river and the 1st came under enemy fire from the opposite bank. Several carbines were capped several times (six or seven) without the cap setting off the cartridge. Due to the defects in their Smith carbines and the ammunition, the 1st Massachusetts suffered one captain and three enlisted men killed. Because of these problems, Colonel Sargent, the commanding officer of the 1st Massachusetts, requested permission on January 22, 1863, to turn in his 508 Smiths and trust his future to the 84 Sharps carbines in his command and the revolver. Three days later, Lieutenant Colonel Alexander Webb, Assistant Inspector General for the Army of the Potomac, was directed to investigate Colonel Sargent's charges. In his report of January 26, Webb stated that he first inspected 84 Smiths at random and found that they were in foul condition at the nipple and breech area. When he questioned the men, most stated that it took several percussion caps to ignite the cartridge. Webb felt the cause of this problem was that most of the Smith cartridges had been in the cavalrymen's cartridge boxes for an extended length of time. While in the cartridge box, most of the powder from the cartridges had leaked out of the holes in the bottom of the cartridges.

Webb next ordered six men from each company, with the cleanest carbines, to form for inspection. When the 42 men formed for inspection, each fired their Smith carbines to test for misfires. The following chart shows the results of this test:[20]

Carbine Number	# of Caps Needed	Carbine Number	# of Caps Needed
1	1	22	1
2	1	23	2
3	2	24	5
4	4	25	6+
5	1	26	1
6	4	27	4
7	3	28	4
8	3	29	2
9	3	30	3
10	3	31	2
11	2	32	4
12	1	33	2
13	2	34	1
14	out of order	35	1
15	2	36	1
16	2	37	4
17	3	38	1
18	2	39	1
19	2	40	2
20	7+	41	2
21	3	42	2

As the preceeding chart reflects, only a quarter of the carbines fired on the first cap and 17 of the 42 required three or more percussion caps to ignite the cartridge. Because of these results, Webb recommended that all Smiths in the regiment be turned in at once. As of April, the 1st Massachusetts was still listing 483 Smiths and 128 Sharps carbines on hand. By Gettysburg, they had been totally rearmed with Sharps carbines.[21]

Spencer

The first Spencer carbines did not reach the field until late in the year. As of the end of December, over 1,000 were listed in field service. The majority of the Spencers were in the following regiments: 2nd New Jersey — 417; 11th Ohio — 368; and 1st West Virginia 177.[22] The field reports for the fourth quarter of the year indicated a favorable opinion for the Spencer. Captain H.W. Dodge, Co. C, 5th Michigan rated the Spencer as very durable and overall a good arm while Captain W.H. Rolls, Co. D, also rated it as a very good arm. Captain Dan Farger, Co. F, 2nd Michigan (Phil Sheridan's first wartime command) stated that the seven-shot Spencers were the best carbines in combat use.[23]

1863 CAVALRY BATTLES
Grierson's Raid

In the western Theater of operations, Grant would call Grierson's raid "one of the most brilliant cavalry exploits of the war." On April 17, Colonel Benjamin Grierson left La Grange, Tennessee, with a force of 1,700 cavalry consisting of the 6th and 7th Illinois and the 2nd Iowa. In 16 days, his raiders rode over 600 miles through Mississippi and, on May 2, reentered the Union lines at Baton Rouge, Louisiana. In this operation, Grierson tied up most of the Confederate cavalry, a third of the infantry and at least two regiments of artillery. This allowed Grant to cross the Mississippi River, unopposed, south of Vicksburg and led to its capture on July 4. In the raid, Grierson suffered 36 casualties while killing and wounding about 100 and capturing 500. Grant also captured 1,000 horses and mules and caused the destruction of large quantities of Confederate supplies and government property.[24]

In June, shortly after the raid, Grierson's cavalry was listing the following small arms in their inventory:[25]

Benjamin H. Grierson, Colonel of the 6th Illinois Cavalry, led his regiment plus the 7th Illinois and 2nd Iowa Cavalry on his raid into Mississippi in April 1863.
U.S. Military History Institute, MOLLUS

6th Illinois:	Cosmopolitan Carbines	436
	Sharps Carbines	147
7th Illinois:	Sharps Carbines	474
	Smith Carbines	94
2nd Iowa:	M1855 Colt Rev. Rifles	313
	Sharps Carbines	238

Brandy Station

At first light on June 9, Buford's cavalry division crossed the Rappahannock in the vicinity of Brandy Station and quickly made contact with Confederate pickets. For the next seven hours, Buford engaged a large part of Jeb Stuart's cavalry. While this battle was in progress, Gregg and Duffie's divisions of cavalry crossed the Rappahannock at about 8:00 a.m. and, by 11:30 a.m., Gregg was nearing Fleetwood Hill. Here, the classic cavalry battle of the war was fought between two of Gregg's brig-

Brigadier General John Buford led the Union Cavalry at Gettysburg on the first day.
U.S. Military History Institute, MOLLUS

ades and Hampton and Jones' Confederate cavalry. After repeated charges and counter charges, the Union cavalry broke off the action late in the afternoon and recrossed the river. Duffie's troops saw little action during the day. The losses on both sides were extensive with the Union listing total casualties of 866 while Jeb Stuart's losses exceeded 500.

At Brandy Station, Buford and Gregg's cavalry were armed with the following small arms:[26]

Brig. Gen. John Buford
8th Ill.	Sharps, Colt .44
3rd Ill.	Sharps, Gallagers, Colt .44
8th N.Y.	Sharps, Colt .44
6th N.Y.	Sharps, Colt .36 and .44
9th N.Y.	S&H, Sharps, Colt .44
17th Penn.	Smith, Merrill, Remington .36
3rd W.V.	Gallager, Colt .44 and Remington .36
6th Penn.	Sharps, Colt .44
1st U.S.	Sharps, Colt .44
2nd U.S.	Sharps, Colt .44
5th U.S.	Sharps, Colt .44
6th U.S.	Sharps, Colt .36 and .44

Brig. Gen. David McMurtrie Gregg
1st Maine	Sharps, Burnside, Colt .44
2nd N.Y.	Sharps, Colt, Remington .36 and .44
10th N.Y.	Smith, Colt .36 and .44
1st Md.	Sharps, Colt .44
1st N.J.	Burnside, Smith, Colt .44
1st Penn.	Sharps, Burnside, Colt .36 and .44

In Jeb Stuart's official report of the battle, he lists the capture of a quantity of ordnance stores as reflected by the chart below:

Captured Ordnance Stores[27] in the Battle of Brandy Station by Cavalry Division, Army of Northern Virginia			
Command	**Sharps Carbines**	**Pistols**	**Sabers**
Hampton	82	64	35
Robertson	2	—	—
Fritz Lee	6	3	5
W.H.F. Lee	15	9	9
W.E. Jones	52	152	107
Horse Artillery	8	4	8
Total	**165**	**232**	**164**

The Turning Point, 1863

Brigadier General David McMurtrie Gregg and staff during the Gettysburg campaign. It was his division and Custer's brigade that engaged Jeb Stuart's cavalry on the third day.
U.S Military History Institute, MOLLUS

Gettysburg

The major battle fought in the East in 1863 was the three-day Battle of Gettysburg from July 1 through 3. The first shots of the battle were by Union pickets located on the Chambersburg Pike at about 7:30 a.m. This post was manned by Company E, 8th Illinois Cavalry with Lt. Marcellus Jones firing the first shot from Sergeant Levi Shafer's Sharps carbine.[28]

July 1

Buford placed the brigades of Devin and Gamble along the Willoughby Run with artillery support on McPherson Ridge. Here, at about 8:00 a.m., the Union cavalry was attacked by two brigades of A.P. Hill's Corps. For over two hours, Buford's troopers were able to delay the Confederate advance on Gettysburg. Finally, the cavalry was relieved by Reynold's First Army Corps of Infantry. Later in the afternoon, Gamble's brigade helped cover the retreat to Cemetery Ridge. The cavalry saw little action on the second day except for skirmishing with Confederate infantry near Culp's Hill.

July 3

At about 2:00 p.m., Jeb Stuart's cavalry appeared about three miles east of Gettysburg where his advance was met by Union cavalry. Here the brigades of McIntosh and Custer fought both mounted and on the skirmish line with Stuart's cavalry. After a hotly contested battle, the Confederates withdrew.

While the cavalry battle was raging east of town, Colonel John Gregg's brigade and Wesley Merritt's regulars were engaged with Confederate infantry in the area of Round Top. During this action, Kilpatrick ordered Elon Farnsworth, a brigadier general of four days, to attack the Confederate position with his cavalry brigade. In this attack, Farnsworth was killed. Of the 300 men making this charge, 65 were casualties. Nearby, at Fairview, the 6th U.S. Regulars were attacked by a large Confederate cavalry force and suffered over 240 casualties of which 203 were prisoners. In the three days of fighting, the Union cavalry suffered 87 killed, 335 wounded and 408 captured or missing in action.

The schedule on the following page reflects the carbines and rifles in the Cavalry Corps of the Army of the Potomac at the time of the battle.[29]

In Grierson's raid through Mississippi in April, the 6th Illinois cavalry was armed with the Cosmopolitan carbine. As of June 30, the regiment was inventorying 436 Cosmopolitans.
U.S. Military Academy collection

Brigade Commander and Unit	CARBINES					RIFLES
	Sharps	Burnside	Smith	Gallager	Merrill	Spencer
Farnsworth						
5th N.Y.	39					
18th Penn.		337				
1st Vt.	140					
1st W.V.		2	81			
Gamble						
8th Ill.	311					
12th Ill.		86				
3rd Ind.	12			182		
8th N.Y.	210					
Devin						
6th N.Y.	232					
9th N.Y.	381		1			
17th Penn.			127		108	
3rd W.V.				89		
McIntosh						
1st Md.	93	8				
Purnell Legion					100	
1st N.J.		128	7			
3rd Penn.	264					
1st Mass.	278					
1st Penn.	298	7				
Custer						
1st Mich.	213	38				
5th Mich.						479
6th Mich.		251				93
7th Mich.		178				
J.I. Gregg						
1st Maine	101	63				
10th N.Y.			93			
4th Penn.	165					
16th Penn.	317					
Merritt						
6th Penn.	231					
1st U.S.	361					
2nd U.S.	245					
5th U.S.	373					
6th U.S.	367					
Huey						
2nd N.Y.	166					
4th N.Y.	245	33				
6th Ohio	208	263				
8th Penn.	50					

The total quantities of small arms listed for the cavalry at Gettysburg were as follows:

Sharps Carbines 4,724
(67 of this # were Sharps and Hankins in 9th N.Y.)
Burnside Carbines 1,387
Smith Carbines 309
Gallager Carbines 271
Merrill Carbines 208
Spencer Rifles 572
*(issued to 5th Michigan and
Co. E and H of the 6th Michigan)*

Morgan's Raid

On July 2, Confederate raider John H. Morgan crossed over the Cumberland River into Kentucky. During the next 24 days, he raised confusion and anxiety in the border states of Kentucky, Indiana and Ohio. Morgan's end came with his capture near Lisbon, Ohio.

During Morgan's raid, the Governor of Indiana telegraphed the Secretary of War for 1,000 carbines

(right) Colonel Alfred Gibbs, 19th New York Cavalry, was issued the Model 1862 Joslyn in September. In his October 4 report, Gibbs gave the Joslyn an overall satisfactory rating but did state that the breechblock (piston) was highly tempered and prone to breakage.
U.S. Military History Institute, MOLLUS

(below) Members of the 16th Pennsylvania Cavalry in the fall of 1863. Note the Pennsylvania state flag that they carried.
National Archives collection

for the state troops. While the Federal government promised help, it never came. The State of Indiana was, therefore, forced to the civilian market where 760 Wesson carbines were purchased from the Cincinnati firm of Kittredge & Co. at a cost of $18,811.40.[30] It does not appear that the Wessons were ever issued for the state's defense. As late as 1869, the state still had in its inventory stores at the Federal arsenal in Indianapolis, 716 Wesson carbines along with 2,093 Springfield rifle muskets and 503 Enfield rifles.[31]

The following list includes the cavalry regiments assigned to hunt down Morgan and the small arms carried in these units:[32]

14th Illinois	Burnside
9th Michigan	Spencer Rifle
5th Indiana	Smith
2nd Ohio	Burnside
1st Kentucky	Burnside & Gallager
4th U.S.	Sharps, Smith & Burnside
8th Michigan	Spencer Rifle

Chickamauga

The largest battle in the West during 1863 occurred in September at Chickamauga. The lion's share of cavalry credits in this battle went to Wilder's mounted infantry brigade armed with Spencer rifles. However, other cavalry regiments besides Wilder's command saw extensive action. They were Colonels Minty, Long and Watkins' brigades, which suffered combined casualties of 28 killed, 120 wounded and 282 missing.[33]

At 6:10 a.m. on September 18, Colonel Minty sent forward scouting patrols looking for the advance elements of Bragg's infantry. Within the hour, he was notified that the Confederates were rapidly approaching in force. For the balance of the day, Minty's 973 troopers, plus two regiments from Wilder's command, held back the Southern forces at Reed Bridge on the Chickamauga River. Late in the day, Minty broke off the action after being informed of a Confederate crossing in force a short distance away.

During the main battle of the September 20, at about 11:00 a.m., Confederate infantry, cavalry and sharpshooters attacked the position held by Colonel Eli Long's cavalry brigade. In this action, Long's force was driven back 200 yards before being able to stop this Southern advance. Shortly after this attack, Long was directed to fall back. In this action, Long suffered 136 casualties out of 900 engaged.

Later in the day, Colonel Louis Watkins' brigade of Kentucky cavalry was attacked on their flank by Confederate infantry.[34] Being in mountainous terrain, Watkins' command lacked mobility and suffered 236 captured. These three Union cavalry brigades were armed as follows:

Colonel Eli Long
2nd Ky.	Merrill, Smith
1st Ohio	Sharps
3rd Ohio	Sharps, Burnside, Merrill
4th Ohio	Sharps, Burnside

Colonel Louis Watkins
4th Ky.	Smith
5th Ky.	Burnside
6th Ky.	Ballard, Burnside

Colonel Robert H.G. Minty
3rd Ind.	Sharps, Smith, Gallager, Burnside
4th Mich.	Colt Revolving Rifles
7th Penn.	Burnside, Sharps, Smith
4th U.S.	Sharps, Smith, Burnside

Company I, 6th Pennsylvania Cavalry (Rush's Lancers) shown in June 1863. By the end of that month, it had turned in the lance for the Sharps carbine.
U.S. Military History Institute, MOLLUS

REGIMENTAL INVENTORIES[35]

When the account was taken of the carbines in field service as of September 30, over 45,000 carbines were listed. As previously stated, not all regiments reported their inventory and others provided only partial listings. This listing gives a good breakdown of the types and quantities of carbines actually in use. First the summary listing and then the actual breakdown:

Carbines In Field Service — September 1863

Sharps 20,378	Merrill 2,953	Ballard 54
Sharps & Hankins . . 123	Hall 1,447	Gibbs 630
Smith 5,105	Gallager 3,670	Joslyn 520
Greene 17	Cosmopolitan . . . 1,535	Pistol Carbines 115
Burnside 8,179	Starr 259	Musketoon 234
Lindner 297	Maynard 78	

Regiment	Count	Type
1st U.S.	78	Sharps
2nd U.S.	106	Sharps
3rd U.S.	390	Sharps
4th U.S.	166	Sharps
	20	Smith
	588	Burnside
5th U.S.	249	Sharps
6th U.S.	436	Sharps
1st Ala. (U.S.)	72	Burnside
	465	Sharps
2nd Ark.	82	Sharps
	14	Merrill
1st Calif.	671	Sharps
2nd Calif.	664	Sharps
1st Batt'l Native Cavalry at San Francisco	72	Sharps
1st Colo.	1	Sharps
1st Conn.	217	Smith
1st Dakota	95	Hall
	75	Sharps
1st Dela.	60	Merrill
2nd Ill.	442	Sharps
3rd Ill.	239	Burnside
	181	Hall
4th Ill.	530	Sharps
5th Ill.	327	Cosmopolitan
	145	Sharps
6th Ill.	130	Cosmopolitan
	369	Sharps
7th Ill.	397	Sharps
	61	Smith
8th Ill.	360	Sharps
9th Ill.	61	Hall
	194	Sharps
	1	Burnside
	1	Smith
10th Ill.	25	Hall
	391	Sharps
11th Ill.	430	Smith
12th Ill.	153	Burnside
13th Ill.	98	Gallager
14th Ill.	173	Burnside
15th Ill.	333	Sharps
16th Ill.	95	Merrill
	100	Sharps
1st Ind.	101	Hall
	10	Sharps
	72	Pistol Carbine
2nd Ind.	384	Smith
3rd Ind.	197	Sharps
	10	Smith
	78	Burnside
	118	Gallager
4th Ind.	209	Smith
5th Ind.	659	Smith
6th Ind.	720	Burnside
65th Ind. Mtd. Inf.	72	Burnside
	8	Smith
1st Iowa	574	Sharps
2nd Iowa	244	Sharps
3rd Iowa	145	Sharps
	3	Hall
4th Iowa	156	Cosmopolitan
	2	Pistol Carbine
5th Iowa	462	Sharps
7th Iowa	640	Gallager
2nd Kan.	75	Merrill
	366	Sharps
5th Kan.	21	Maynard
	481	Sharps
	196	Starr
6th Kan.	55	Hall
	253	Merrill
	7	Maynard
	253	Sharps
	13	Pistol Carbine
7th Kan.	134	Sharps
	4	Maynard
9th Kan.	549	Gallager
	14	Greene
	2	Sharps
11th Kan.	80	Hall
	81	Sharps
14th Kan.	62	Merrill
	53	Sharps
1st Ky.	33	Burnside
	15	Gallager
2nd Ky.	197	Merrill
	72	Smith
3rd Ky.	30	Ballard
	270	Smith
4th Ky.	28	Smith
5th Ky.	99	Burnside
6th Ky.	54	Ballard
	19	Burnside
7th Ky.	52	Burnside
	40	Smith
8th Ky.	66	Gallager
11th Ky.	106	Burnside
14th Ky.	50	Cosmopolitan
15th Ky.	293	Gallager
11th Ky. Mtd. Inf.	262	Smith
1st La.	267	Burnside
	38	Sharps

Continued on page 36

Unit	Qty	Carbine
1st Maine	68	Burnside
	112	Sharps
1st Mass.	165	Sharps
2nd Mass.	43	Burnside
	65	Sharps
3rd Mass.	378	Sharps
1st Md.	21	Burnside
	89	Smith
2nd Md.	935	Smith
Purnell Legion	111	Merrill
Potomac Home Brig.	278	Sharps
1st Mich.	20	Burnside
	157	Sharps
2nd Mich.	3	Sharps
3rd Mich.	92	Burnside
	1	Sharps
	2	Smith
6th Mich.	134	Burnside
7th Mich.	106	Burnside
1st Minn.	236	Sharps
	571	Smith
Minn. Independent Batt'l	200	Burnside
1st Mo.	131	Sharps
2nd Mo.	246	Hall
	239	Sharps
3rd Mo.	317	Sharps
4th Mo.	130	Gallager
	12	Hall
6th Mo.	98	Burnside
	2	Cosmopolitan
	23	Hall
	290	Sharps
7th Mo.	192	Hall
	308	Sharps
8th Mo.	433	Cosmopolitan
10th Mo.	426	Gibb
	1	Hall
11th Mo.	512	Merrill
	147	Sharps
1st Mo. State Militia	60	Burnside
3rd Mo. State Militia	251	Hall
	58	Sharps
	25	Musketoon
4th Mo. State Militia	81	Gallager
	119	Hall
5th Mo. State Militia	25	Pistol Carbine
	55	Musketoon
6th Mo. State Militia	47	Sharps
1st Miss.	192	Sharps
1st N.J.	218	Burnside
	13	Sharps
	7	Smith
1st N.M.	6	Sharps
	154	Musketoon
1st Nev.	299	Sharps
1st N.Y.	97	Burnside
	77	Gallager
	22	Sharps
2nd N.Y.	113	Sharps
	2	Starr
3rd N.Y.	282	Sharps
	86	Burnside
4th N.Y.	5	Burnside
	177	Sharps
5th N.Y.	25	Sharps
6th N.Y.	5	Burnside
	299	Sharps
7th N.Y.	546	Sharps
8th N.Y.	298	Sharps
9th N.Y.	2	Burnside
	253	Sharps
	(S&H = 123)	
	1	Smith
10th N.Y.	5	Sharps
	135	Smith
11th N.Y.	376	Burnside
12th N.Y.	352	Burnside
	50	Starr
13th N.Y.	93	Gibbs
14th N.Y.	41	Burnside
16th N.Y.	111	Gibbs
18th N.Y.	51	Smith
19th N.Y.	520	Joslyn
23rd N.Y.	4	Sharps
1st Ohio	288	Sharps
2nd Ohio	218	Burnside
3rd Ohio	76	Burnside
	11	Merrill
	153	Sharps
4th Ohio	57	Burnside
	293	Sharps
5th Ohio	354	Burnside
	101	Sharps
6th Ohio	137	Burnside
	8	Sharps
	1	Smith
7th Ohio	177	Burnside
	45	Sharp
9th Ohio	166	Gallager
10th Ohio	9	Gallager
	80	Merrill
	169	Sharps
11th Ohio	3	Greene
	2	Hall
	3	Sharps
4th Ohio Batt'l	54	Burnside
5th Ohio Batt'l	377	Cosmopolitan
1st Penn.	23	Burnside
	270	Sharps
2nd Penn.	391	Sharps
3rd Penn.	11	Burnside
	258	Sharps
4th Penn.	166	Sharps
5th Penn.	239	Sharps
6th Penn.	31	Sharps
7th Penn.	22	Sharps
	115	Burnside
	242	Smith
8th Penn.	107	Sharps
	16	Burnside
9th Penn.	420	Burnside
	51	Sharps
11th Penn.	80	Sharps
12th Penn.	146	Burnside
13th Penn.	298	Burnside
	19	Sharps
	7	Gallager
	4	Starr
14th Penn.	471	Burnside
15th Penn.	67	Burnside
	384	Sharps
16th Penn.	166	Sharps
17th Penn.	2	Burnside
	44	Merrill
	189	Sharps
	23	Smith
18th Penn.	283	Burnside
20th Penn.	148	Gallager
21th Penn.	419	Gallager
22th Penn.	47	Gallager
	95	Sharps
Dana's Independent Co. Cav.	82	Sharps
1st R.I.	136	Sharps
1st Tenn.	191	Gallager
2nd Tenn.	70	Burnside
	228	Merrill
3rd Tenn.	113	Merrill
	328	Sharps

4th Tenn.	523 Gallager	1st Texas	141 Sharps		1 Sharps
5th Tenn.	305 Merrill	1st Vt.	102 Sharps		70 Smith
	117 Smith		9 Starr	7th W.V.	297 Lindner
6th Tenn.	452 Sharps	1st W.V.	45 Sharps	1st Wisc.	303 Merrill
7th Tenn.	110 Cosmopolitan		3 Smith		46 Maynard
	127 Sharps	2nd W.V.	4 Sharps		15 Sharps
10th Tenn.	87 Merrill		187 Smith	2nd Wisc.	497 Sharps
1st Independent Co. Cav.		3rd W.V.	63 Burnside	3rd Wisc.	403 Merrill
	60 Burnside		93 Gallager	4th Wisc.	320 Burnside

Footnotes — The Turning Point, 1863

[1] ORs Vol. 27, Series I, Part II, pp. 206 and 209
[2] NARG 156-110
[3] ORs Vol. 29, Series I, Part I, p. 41
[4] *Ibid.,* p. 503
[5] NARG 156-21, Box 283
[6] *Ibid.,* Box 275
[7] *Ibid.,* Box 283
[8] ORs Vol. 22, Series I, Part I, pp. 391 and 402-404
[9] NARG 108-75
[10] NARG 156-1001
[11] NARG 156-215
[12] NARG 108-75
[13] *Ibid.*
[14] NARG 156-6 and 215
[15] NARG 156-215
[16] NARG 108-75
[17] NARG 156-1001
[18] NARG 156-215 and 108-75
[19] NARG 156-21, Box 223
[20] NARG 156-1001
[21] NARG 156-110, March 31, 1863
[22] *Ibid.,* December 31, 1863
[23] NARG 156-215
[24] ORs Vol. 24, Series I, Part I, pp. 522-529
[25] NARG 156-110
[26] *Ibid.*
[27] ORs Vol. 27, Series I, Part II, p. 719
[28] Abner Hard, M.D., *History of the Eighth Cavalry Regiment, Illinois Volunteers,* Dayton, OH, 1984, p. 256
[29] NARG 156-110
[30] State of Indiana Adjutant General Report of 1869, Indianapolis, IN, p. 436
[31] *Ibid.,* p. 442
[32] NARG 156-110 and ORs Vol. 23, Series I, Part I, p. 637
[33] *Battles and Leaders of the Civil War, Retreat from Gettysburg, Vol. III,* New York, 1956, p. 673
[34] ORs Vol. 30, Series I, Part I, pp. 914-923
[35] NARG 156-110

A Union cavalryman shown with his Burnside carbine.

The Herb Peck Jr. collection

THE YEAR OF ATTRITION
◆ 1864 ◆

Late in the previous year, the Spencer and Starr carbines had reached the troops in the field. In 1864, these arms would be fielded in even larger quantities. While the Spencer was to receive rave reviews, the Starr was nearly always condemned as totally worthless. This year was the initial fielding of the Warner and the war production model Maynard carbines; 1864 also saw continual acceptance for the Sharps and the death of the young colonel of the 9th New York Cavalry (Colonel William Sackett) who had advocated the merits of the Sharps & Hankins so strongly.

The Federal arsenals and temporary depots as of Saturday, January 2, had in storage over 18,000 carbines. Eleven months later, on November 5, this total had been reduced to about 8,000 carbines. The quantities in storage are shown in the following schedule:

Carbines In Storage[1]
As of January 2 & November 5, 1864

	Jan. 2	Nov. 5		Jan. 2	Nov. 5
Ballard	126	501	Sharps	630	1,006
Burnside	4,019	296	Sharps & Hankins		37
Cosmopolitan	715	12	Smith	1,614	1,058
Joslyn	755	335	Starr	1,220	701
Gallager	2,263	169	Spencer	3,245	647
Gibbs	168	218	Warner		549
Greene	2	11	Pistol Carbine	559	178
Hall	639	320	Rifled Musketoon	166	45
Lindner		344	Smooth-bore		
Merrill	922	619	Musketoon	1,407	125
Maynard	22	1,191	Wesson		286

During the year, the Ordnance Department had replaced 93,394 carbines that were either damaged, destroyed or lost in battle.

1864 FIELD REPORTS

Spencer

By September, over 9,000 Spencer carbines were listed in 38 regiments. From the field, came reports rating the Spencer as the best carbine in the service. This opinion is reflected in the following reports: In January, Lieutenant Colonel Gould, 5th Michigan Cavalry, stated that "the Spencer carbine is preferred instead of the Spencer rifle as a superi-

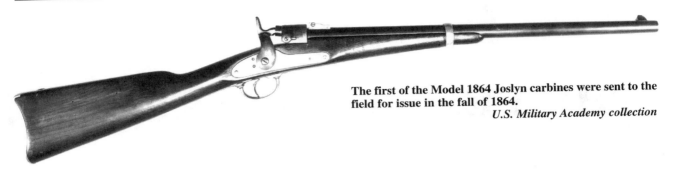

The first of the Model 1864 Joslyn carbines were sent to the field for issue in the fall of 1864.
U.S. Military Academy collection

(above) The most advanced carbine to see action during the war was the seven-shot .56-56 Spencer carbine. Shown here is a close-up of the action of the Spencer.
Author's collection

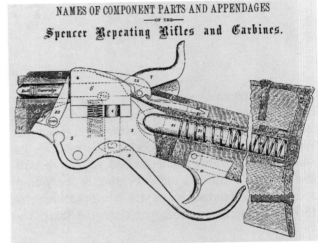

(right) The top portion of a parts list for the Spencer provided by the company to the Ordnance Department. This cross section drawing shows the action of the seven-shot Spencer.
National Archives collection

Nearly half of the Union cavalry in 1864 were armed with Sharps carbines. *The Herb Peck Jr. collection*

or arm. It is less apt to break and more easily carried."[2] In June, ten officers of the 4th U.S. Cavalry rated the Spencer the best arm in cavalry service. The 4th was the only regiment of regulars during the war to have been issued Spencer carbines.[3]

Major Gorke of the 2nd New Jersey, whose regiment was armed with 379 Spencers, stated on March 2 that the range, accuracy and rate of fire should recommend it to the attention of all officers. The lieutenant colonel of the 4th Iowa with 669 Spencer carbines regarded them as a very fine cavalry arm. Captain Phipps of the 16th Pennsylvania Cavalry felt that the Spencer was the most effective weapon in the cavalry and that at least two regiments in each brigade should be armed with the Spencer carbine.[4]

Sharps

Hundreds of reports regarding the Sharps carbine were received throughout the war. The vast majority of these reports were listing it as a very durable arm. The following reports are examples of the 1864 responses from the field. In January, Lieu-

The Year of Attrition, 1864

New Model 1863 Sharps carbine sn. 78,318 was issued to Company B, 1st California Cavalry in 1864. Company B was one of the units with Kit Carson in the Adobe Walls fight on November 25. *Author's collection*

In November 1864, when this photo of the 3rd Indiana Cavalry was taken, the regiment was armed with Burnsides. *U.S. Military History Institute, MOLLUS*

tenant Colonel Wallace of the 4th Illinois rated the Sharps as the best arm that he had observed. These same comments were received from Major Seley, 5th Illinois; Colonel Stephens, 2nd Wisconsin; Colonel Hovick, 7th Kansas; Lieutenant Colonel Caldwell, 1st Iowa; Major Walsh, 3rd Tennessee; Captain Preston, Merrill's Horse (2nd Missouri); and Major Crowninshield, 2nd Massachusetts whose regiment was armed, in January, with 557 Sharps.[5]

Major Thompson, 1st Iowa, Chief of Staff for the 7th Army Corps in the Department of Arkansas states: "Sharps carbines have no superior. In range it is fully equal to a minie rifle and the facility and accuracy with which it is loaded and discharged renders our dismounted cavalry more than equal the enemy's infantry and far superior to their cavalry."

One area of concern was raised by Major Patton, 3rd Indiana Cavalry. The problem was that the small springs in the lock were easily broken, but an armorer could easily repair most of the carbines if the springs were made available. At the time of Major Patton's report in January, his regiment was equipped with 247 Sharps carbines.[6]

Ballard

By the end of March, nine reports had been received on the merits of the Ballard carbine. These reports had come from the 11th and 13th Kentucky Cavalry and the 45th Kentucky Mounted Infantry. The officers of the 13th considered it as a first-rate cavalry arm. These officers consisted of Colonel J.W. Westherford, Captains Asa Bayant of Co. B, Howard of Co. F, Crandell of Co. G, Northrup, Co. H and Penn, Co. M. These officers did state that the spring that extracts the cartridge case had a tendency to break. Colonel Brown of the 45th liked the Ballard but did not care for the swivel attachment for the sling. In April, Colonel Brown's 45th Kentucky was armed with 687 Ballard carbines. The 11th Kentucky felt it carried extremely well and was a splendid carbine.[7]

Major W.H. Fidler of the 6th Kentucky Cavalry refused to turn in his 150 Ballards when the regiment was mustered in as veterans, fearing that he could not obtain a more effective carbine for cavalry use.[8] The major felt that the Ballard was the best carbine in the service.

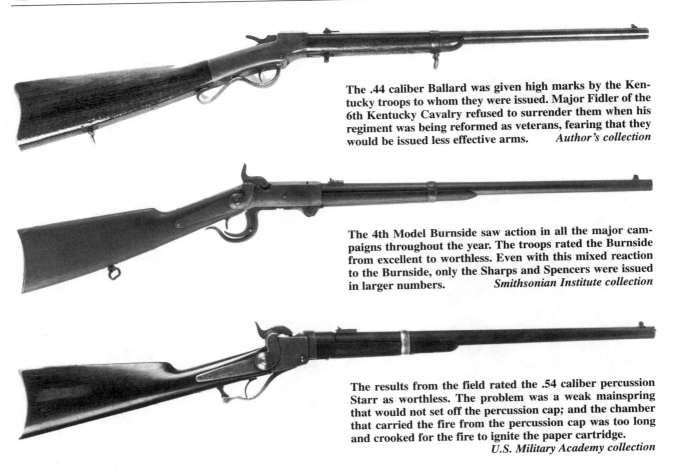

The .44 caliber Ballard was given high marks by the Kentucky troops to whom they were issued. Major Fidler of the 6th Kentucky Cavalry refused to surrender them when his regiment was being reformed as veterans, fearing that they would be issued less effective arms. *Author's collection*

The 4th Model Burnside saw action in all the major campaigns throughout the year. The troops rated the Burnside from excellent to worthless. Even with this mixed reaction to the Burnside, only the Sharps and Spencers were issued in larger numbers. *Smithsonian Institute collection*

The results from the field rated the .54 caliber percussion Starr as worthless. The problem was a weak mainspring that would not set off the percussion cap; and the chamber that carried the fire from the percussion cap was too long and crooked for the fire to ignite the paper cartridge. *U.S. Military Academy collection*

Burnside

The Burnside field reports covered the full range of opinions from an excellent arm to totally unreliable. Here are examples of these diverse opinions.

In April, Colonel Boardman, 4th Wisconsin Cavalry, with 422 Burnsides, stated that they were continually getting out of order because of the complexity of their parts. Colonel Savage of the 12th New York with 510 Burnsides and 83 Starrs rated the Burnside superior to the Starr carbine, while Captain Snider, 10th Indiana, considered the Burnside a good single-shot carbine.

Colonel Reno of the 12th Pennsylvania, with nearly 700 Burnsides, considered them the least effective arm in the cavalry. Reno states its range and accuracy was poor, the ammunition was heavy and, when exhausted, the carbine was worthless — plus it often misfired. Major Douglas of the 20th Pennsylvania stated that many of the barrels burst near the muzzle and the cartridges generally required the bullet to be cut in order to get them into the chamber.[9]

Starr

The results from the field on the Starr carbine were basically negative. Colonel Cesnola of the 4th New York rated the Starr of little or no use. Colonel Trumbull of the 9th Iowa with 908 Starrs stated in March that the sights came loose and it misfired badly, while Lieutenant Colonel Moyer of the 3rd Michigan with 894 Starr carbines reported in May that the carbine worked miserably with the caps furnished. In July, Captain Preston from Merrill's Horse (2nd Missouri) with 339 Starrs said that they frequently misfired with the same ammunition used with the Sharps. This same comment was made by Captain James Green of the 12th Missouri Vol. Cavalry who stated that they hung fire and often took three or four caps to ignite the paper cartridge. The cause of this problem may be found in the report from a lieutenant colonel in a Tennessee cavalry regiment that stated that their defects were mostly in the weak mainspring in the lock which caused nine out of ten carbines to fail to snap a cap; also the tube to carry fire to the cartridge was too long and the passage too crooked.[10]

A cavalryman with his Ballard carbine.
The Herb Peck Jr. collection

Gallager

From the officers in the field came reports on the Gallager carbine ranging from utterly worthless to a very good arm. The one cavalry regiment that gave the Gallager a very good rating was the 9th Kansas. In September, the 9th was inventorying 444 Gallagers and nine pistol carbines.[11] An opposite opinion was received from Major Dyer, 1st Tennessee, who stated that their Gallagers had been condemned as worthless and were replaced with Burnsides.

Problems raised by these reports were that some of the Gallager guards broke while being fired; and

Members of the 9th Pennsylvania Cavalry at Lookout Mountain shown with their percussion Starr carbines.
U.S. Military History Institute

Lieutenant Colonel Hartmann, 13th Illinois, stated that the brass cover to the cartridge delayed loading by sticking to the barrel. The difficulty in withdrawing the cartridge case after a few firings was raised by Major Langen, 4th Missouri. At the time of Langen's report, his regiment was armed with 301 Gallagers. Colonel Phelps, 2nd Arkansas, rated the Gallager less effective than the Sharps, Smith or Cosmopolitan. He also felt they were less efficient in their point of range, accuracy and durability. Lieutenant Colonel Prosser of the 2nd Tennessee, with 589 Gallagers, complained in July of the Gallager being inadequate compared to the Confederate Enfield rifles.

Cosmopolitan (Union or Gwyn & Campbell)

One regiment armed with the Cosmopolitan carbines throughout 1864 was the 8th Iowa Cavalry. During the first three months of the year, the regiment was stationed near Nashville, Tennessee where they operated against guerrillas. From April through September, the 8th was part of Sherman's cavalry in the Atlanta campaign and they closed the year back at Nashville.

Prior to the Atlanta Campaign, the officers from the 8th Iowa had rated the Cosmopolitan as a very effective carbine. As of March, they were listing an inventory of 966 Cosmopolitans on hand. By August, their opinion had totally changed. Major Price, Co. E, stated, "I find that they are not worth a damm, have lost men in the campaign by the worthlessness of the arm." Captain Evans, Co. B, claimed, "It is impossible to keep the chamber dry." Lieutenant Jacob Hardin, Co. D, reported that "they nearly put a man's eyes out everytime they fire them, the powder burns their faces." As of August, the 8th rated the Cosmopolitan as worthless for cavalry use.[12]

A second regiment to have similar views of the merits of the Cosmopolitan was the 8th Ohio Cavalry. During the latter part of the year, they were operating in the vicinity of Beverly, West Virginia.

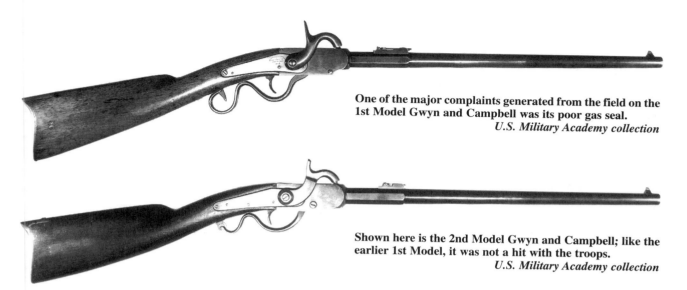

One of the major complaints generated from the field on the 1st Model Gwyn and Campbell was its poor gas seal.
U.S. Military Academy collection

Shown here is the 2nd Model Gwyn and Campbell; like the earlier 1st Model, it was not a hit with the troops.
U.S. Military Academy collection

Captain C.H. Evans, Co. F, had these complaints. It failed to carry up to the sights, leaked fire at the breech, became fouled after a few discharges; the hammer often failed to snap a cap and the least jar broke the carbine at the upper end of the stock. Captain Robert Lyle, Co. M, stated that the Cosmopolitans were utterly inefficient over one hundred yards.

Not all regiments rated the Cosmopolitan as an ineffective arm. The 5th Illinois and 8th Missouri Vol. Cavalry gave it an overall good rating. Major Rich, commanding officer of the 8th Missouri, rated his 401 Cosmopolitan carbines as first-rate arms which was seconded by Major Seley of the 5th Illinois. However, later in the year, the 5th Illinois' 271 Cosmopolitans were condemned and the regiment was totally rearmed with Sharps carbines.

1864 CAMPAIGNS & BATTLES
Yellow Tavern

The cavalry battle at Yellow Tavern on May 11 on the outskirts of Richmond resulted in the defeat of the Southern cavalry and the death of their leader Jeb Stuart. Brigadier General James Wilson commented on Stuart's death: "from it may be dated the permanent superiority of the national cavalry over that of the rebels."[13]

The climax of the battle occurred around four p.m. with Custer's 1st Michigan mounted cavalry charge supported by the 5th and 6th Michigan's dismounted attack. In the attack, the Michigan Brigade broke the Confederate position on the left. The balance of the Confederate positions were attacked and driven back by the brigades of Merritt, Chapman and Devin. In an attempt to rally his men, Jeb Stuart was shot by private John A. Huff of Company E, 5th Michigan Cavalry. Huff, formerly with Berdan's Sharpshooters, shot Stuart with a .44 caliber Colt Model 1860 revolver. Private Huff, himself, was killed a few weeks later at Haw's Shop. At Yellow Tavern, the Union suffered 35 killed, 142 wounded and 82 missing.

Trevilian Station

On June 5, Grant issued orders to Sheridan to make a raid toward Charlottesville and, if possible, to destroy the railroad bridge over the Rivanna. Two days later, Sheridan left with 6,000 officers and men from the divisions of Torbert and Gregg. The men left with three days' rations, two days' grain for the horses and forty rounds of ammunition per man. The reserve ammunition of sixty rounds per cavalryman was carried in the wagons.

By the evening of June 10, the Union cavalry was nearing Trevilian Station. Early the next morning, Wade Hampton's 4,700-man Confederate cavalry attacked Sheridan. During the early stages of the battle, Custer's Michigan Brigade found itself in the Confederate rear in sight of Trevilian Station. Here Custer captured a quantity of supplies and took 800 prisoners, but he was in turn attacked from all sides. Custer was able to hold his ground until the brigades of Merritt, J. Irvin Gregg and Devin came to his assistance. On the afternoon of June 12, the Union cavalry attacked Hampton's cavalry, which had taken up defensive positions a couple of miles west of Trevilian Station. Seven attacks were made on these defensive works without success. Lacking ammunition to continue the fight and with a large number of wounded, Sheridan withdrew to his base of supplies located at White House. In the two days of fighting at Trevilian Station, Sheridan's casualties consisted of 102 killed, 470 wounded and 435 missing. Hampton listed his losses at 612. Fitzhugh Lee's casualties were not reported.[14]

By the end of June, the Union cavalry in Grant's 1864 summer campaign had engaged the Southern forces at Todd's Tavern, Yellow Tavern, Haw's Shop, Cold Harbor and Trevilian Station.

As of June 30, the following arms were listed for the cavalry of the Army of the Potomac:

Sharps Carbines 4,854
Spencer Carbines 1,045
Burnside Carbines 1,942
Smith Carbines 257
Starr Carbines 13
Gallager Carbines 25
Joslyn Carbines 22
Sharps & Hankins Carbines 184
Spencer Rifles 175
Springfield Rifle Muskets 2

The following is a breakdown of arms by division and regiment:

1ST DIVISION — BRIG GEN. ALFRED TORBERT

1st Brigade — Brig. Gen. George Custer

	Sharps	Spencer Carbine	Spencer Rifle	Burnside
1st Michigan	40	264		
5th Michigan	13	102	37	
6th Michgan		244	1	1
7th Michgan		19	66	7
Total	**53**	**629**	**104**	**8**

2nd Brigade — Col. Thomas Devin

	Sharps	Sharps & Hankins	Burnside	Smith
4th New York	182			
6th New York	238			
9th New York	46	184		
17th Pennsylvania	235		2	4
Total	**701**	**184**	**2**	**4**

Reserve Brigade — Brig. Gen. Wesley Merritt

	Sharps	Spencer	Joslyn	Rifle Musket
19th New York		220	22	2
6th Pennsylvania	195			
1st U.S.	129			
2nd U.S.	204			
5th U.S.	89			
Total	**617**	**220**	**22**	**2**

2ND DIVISION — BRIG. GEN. DAVID McMURTRIE GREGG

1st Brigade Brig. Gen. Henry Davies, Jr.

	Sharps	Burnside	Smith	Spencer Carbine
1st Massachusetts	353	47		4
1st New Jersey		288		
10th New York	118	21	104	
6th Ohio	108	100	3	
1st Pennsylvania	217	13		
Total	**796**	**469**	**107**	**4**

2nd Brigade Col. J. Irvin Gregg

	Sharps	Burnside	Gallager	Spencer Carbine
1st Maine	172	186		
2nd Pennsylvania	392			
4th Pennsylvania	203			
8th Pennsylvania	129	62		
13th Pennsylvania	178	141	25	10
16th Pennsylvania	380	12		
Total	**1,454**	**401**	**25**	**10**

3RD DIVISION — BRIG. GEN. JAMES WILSON

1st Brigade Col. John McIntosh

	Sharps	Burnside	Smith	Starr	Spencer Carbine	Rifle
1st Connecticut	25	56	64		14	4
3rd New Jersey	172			2		
2nd New York	137	191		1		42
5th New York	7				152	22
2nd Ohio	39	343		9	1	3
18th Pennsylvania		226				
Total	**380**	**816**	**64**	**12**	**167**	**71**

2nd Brigade Col. George Chapman

	Sharps	Burnside	Smith	Spencer Carbine
3rd Indiana*	133	243	78	
8th New York	140		4	
1st Vermont	165	2		15
Total	**438**	**245**	**82**	**15**

Escort

	Sharps	Starr
6th U.S.	317	
8th Illinois	98	1

* Part of this regiment was in the West and a portion of the above total includes the Western portion of this unit.

Atlanta Campaign

General Sherman opened the Atlanta Campaign on May 5 with a cavalry strength of 12,455 officers and men. With the capture of Atlanta on September 3rd, the Union cavalry consisted of nearly 40 cavalry and mounted infantry regiments from the Armies of the Cumberland, Tennessee and Ohio.[15]

Not until Sherman neared Atlanta did the cavalry play a role in its capture. Sherman had been reluctant to use his cavalry, believing that the Confederate cavalry outnumbered his; he also thought that much of the terrain that the army passed through was totally unsuitable for large scale cavalry action. The probable reason for Sherman's not using his cavalry was a lack of trust in his senior cavalry commanders. In the vicinity of Atlanta, the cavalry made several raids on the Confederate railroads with little success. In one of these raids, General Stoneman was captured. In a separate raid, Kilpatrick's cavalry, consisting of Colonel Minty's and Long's brigades, had to fight its way out of a Confederate encirclement. Kilpatrick suffered 451 casualties or about 10 percent of his force. Kilpatrick had been facing a force of 12,000 infantry, cavalry and Georgia Militia.[16]

During the Atlanta Campaign, Sherman's cavalry and mounted infantry were armed as follows:

Atlanta Campaign
May 5 - September 3, 1864[17]

Army of Tennessee
5th Ohio	Sharps
1st Alabama	Smith, Burnside

Army of Ohio
8th Mich.	Burnside
9th Mich.	Spencer, Burnside
7th Ohio	Burnside, Ballard
14th Ill.	Sharps, Burnside
16th Ill.	Sharps
5th Ind.	Sharps
6th Ind.	Sharps, Gallager, Burnside
1st Ky.	Smith, Burnside
11th Ky.	Smith, Burnside, Ballard
12th Ky.	Ballard
McLaughlin's Squadron	Ballard, Burnside

Army of the Cumberland
92th Ill. Mounted Inf.	Burnside, Spencer Rifle
98th and 123rd Ill. Mtd. Inf.	Spencer Rifle
2nd Mich.	Spencer
4th Mich.	Sharps, Spencer
4th U.S.	Spencer
2nd Ind.	Smith, Merrill, Spencer
3rd Ind.	Burnside
4th Ind.	Smith
8th Ind.	Spencer, Joslyn
17th Ind. Mtd. Inf.	Burnside, Spencer Rifle
72nd Ind. Mtd. Inf.	Spencer Rifle
5th Iowa	Sharps
8th Iowa	Cosmopolitan, Burnside
2nd Ky.	Sharps, Spencer, Joslyn
3rd Ky.	Sharps
4th Ky.	Sharps, Smith, Burnside
5th Ky.	Burnside
6th Ky.	Sharps, Smith, Ballard, Burnside
7th Ky.	Sharps, Spencer, Smith, Burnside
4th Ky. Mtd. Inf.	Merrill, Ballard, Spencer
1st Ohio	Sharps, Burnside
3rd Ohio	Sharps, Spencer, Burnside
4th Ohio	Sharps, Spencer, Burnside
10th Ohio	Sharps
7th Penn.	Sharps, Spencer, Burnside
1st Tenn.	Sharps, Burnside
1st Wisc.	Sharps, Merrill, Joslyn, Burnside

On November 12, Sherman left Atlanta and started his "March to the Sea" that ended with the capture of Savannah on December 21. On Sherman's March, his army consisted of 55,000 infantry and artillery as well as 5,500 cavalry under the command of Judson Kilpatrick. By the end of the "March to the Sea," Kilpatrick's cavalry had expended 56,000 Burnside cartridges, 500 Henry, 62,000 Sharps, 21,000 Smith and 141,396 Spencer rifle cartridges.[18]

1864 Valley Campaign

On August 7, Phil Sheridan took overall command of all the forces in the valley now named the Middle Military Division. His task was to defeat Jubal Early's Confederate forces and then to destroy the valley's vast quantities of food supplies for the Confederate army.

Sheridan's cavalry was under the command of Alfred Torbert whose divisional commanders were Wesley Merritt, William Averell and James Wilson. On the way to the valley from the Petersburg area, Wilson stopped off in Washington and was able to have the McIntosh Brigade, consisting of the 1st Connecticut, 3rd New Jersey, 2nd and 5th New York, 2nd Ohio and 18th Pennsylvania, armed by the Ordnance Department with Spencer carbines.[19] In fact, of the 34 regiments of cavalry assigned to the valley, 18 regiments were armed with Spencer carbines as of the end of September. The total number of Spencers at this time came to nearly 3,400.[20] By the end of the year, this total had increased to over 5,300.[21]

Alfred Torbert, Commanding[22]
Escort — 1st R.I. — Sharps
As of September 1864

First Division, Wesley Merritt
1st Brigade, George Custer
- 1st Mich.   Spencer
- 5th Mich.   Spencer
- 6th Mich.   Spencer
- 7th Mich.   Spencer
- 25th N.Y.   Burnside

2nd Brigade, Thomas Devin
- 4th N.Y.   Sharps
- 6th N.Y.   Sharps
- 9th N.Y.   Sharps, Starr
- 19th N.Y.   Spencer
- 17th Penn. Spencer, Starr

Reserve Brigade, Charles Lowell, Jr.
- 2nd Mass.   Sharps, Burnside
- 6th Penn.   Sharps
- 1st U.S.   Sharps
- 2nd U.S.   Sharps
- 5th U.S.   Sharps

Second Division, William Averell
1st Brigade, James Schoonmaker
- 8th Ohio   Spencer, Burnside
- 14th Penn. Spencer, Burnside
- 22th Penn. Sharps, Smith, Starr, Burnside

2nd Brigade, Henry Capehart
- 1st N.Y.   Sharps
- 1st W.V.   Spencer, Smith
- 2nd W.V.   Spencer, Smith
- 3rd W.V.   Spencer, Smith

Third Division, James Wilson
1st Brigade, John McIntosh
- 1st Conn.   Spencer, Smith
- 3rd N.J.   Spencer, Burnside
- 2nd N.Y.   Spencer
- 5th N.Y.   Spencer
- 2nd Ohio   Spencer
- 18th Penn. Spencer, Burnside

2nd Brigade, George Chapman
- 3rd Ind.   Burnside
- 1st N.H.   Sharps
- 8th N.Y.   Sharps
- 22nd N.Y.   Sharps
- 1st Vt.   Spencer

On September 13, Brigadier General John McIntosh was directed to take his brigade and make a reconnaissance in force towards Winchester, Virginia. Within about two-and-a-half miles from Winchester, McIntosh crossed the Opequon and attacked the Confederate cavalry posted on a hill overlooking the creek, capturing 37 prisoners. Continuing on, he surrounded and captured the entire 8th South Carolina Infantry including 14 officers and 92 enlisted men. The 8th's battle flag was captured by Corporal Issac Gause of Company E, 2nd Ohio Cavalry. For this action, Gause was awarded the Congressional Medal of Honor. McIntosh's casualties consisted of two killed and three wounded.[23]

Six days later, at the Battle of Opequon Creek, the Union cavalry played a major role in the Union

In Philip Sheridan's 1864 Valley Campaign, his cavalry was led by Alfred T.A. Torbert shown here with his staff. *U.S. Military History Institute, MOLLUS*

victory. The brigades of McIntosh and Averell were able to cross the Opequon with little or no resistance. This was not the case with Devin, Lowell and Custer. They had to force their passage across the creek under Confederate fire. Custer dismounted the 6th Michigan, with their Spencer carbines, to act as cover fire for the mounted charge of the 7th Michigan and the 25th New York. When this attack failed, a second, successful attempt was made by the 1st Michigan Cavalry. Colonel Lowell, with the Reserve Brigade, also had to dismount a portion of his command with their Sharps carbines to act as cover for the mounted attack. At about 2:00 p.m., five brigades of Union cavalry attacked the Confederates on their flank and rear with a series of mounted sabre charges that broke the Confederate lines and helped bring about their defeat. In the battle, the Union cavalry suffered casualties of 65 killed, 267 wounded and 109 missing.[24]

The valley campaign lasted for the balance of the year with the cavalry involved in such actions as Fisher's Hill, "Woodstock Races" and Cedar Creek. In this fall campaign, the Union horsemen suffered over 3,900 casualties, which reflects the substantial part played by Sheridan's cavalry.

Adobe Walls

One of the largest Indian battles in U.S. history occurred in November, at Adobe Walls, between a cavalry detachment led by Colonel Christopher "Kit" Carson and the Kiowa and their allies, the Comanche. On November 12, Carson left Fort Bascom, New Mexico with 14 officers and 321 enlisted personnel plus 75 Ute and Apache scouts to attack the Kiowa and Comanches in their winter camps about 200 miles away on the Canadian River. The cavalry consisted of Companies B, K and M of the 1st California and Companies D and M of the 1st New Mexico with a total enlisted strength of 236. The infantry of 95 men were from the 1st California Infantry plus two mountain howitzers commanded by Lieutenant Pettis. The infantry was armed with the .58 caliber rifle musket and the cavalry with Sharps carbines.

Early on the morning of November 25, Carson left his wagon trains guarded by his infantry and moved forward to attack the hostiles with his cavalry and Indian allies. In total, his force consisted of about 300 men. In a surprise attack on a Kiowa village of 150 lodges, the cavalry forced the Indians to retreat downstream a few miles to Adobe Walls where 350 Comanche lodges were located. Here Carson ran into a hornet's nest of hostile Indians. In the day-long engagement, the cavalry, dismounted and fighting on a skirmish line, were able to hold off a combined Indian force of over 1,000 warriors. Only the skilled command of Carson and the effective use of the two mountain howitzers prevented them from being overrun. Late in the day, Carson retreated through the Kiowa village and destroyed it before returning to his wagon train. At the village, the howitzers were again used to keep the Indians at bay. In the day-long fight, Carson listed his casualties as two killed and ten wounded; the Utes suffered one killed and five wounded. Carson listed Kiowa and Comanche losses at sixty.[25]

Confederate Field Service

The Confederate cavalry was issued a wide variety of small arms in 1864 as can be seen in the following three inspection reports. Many of the arms in these reports had been previously captured from the Union cavalry. In November, when Sherman was marching to the sea, part of the Confederate cavalry opposing the Union advance was Brigadier General Lawrence Ross' cavalry brigade armed as shown below:[26]

	3rd Texas	6th Texas	9th Texas	27th Texas
Enfield Rifles	98	83	65	87
Austrian Rifles	12	35	7	10
Sharps Carbines	49	48	32	46
Spencer Carbines	28	21	11	13
54 cal. Ballard Carbines	4		1	4
Union Carbines	3	2	15	5
Burnside Carbines		2	5	5
Gallager Carbines		2		
Smith Carbines		1	4	4
Merrill Carbines				1

At the start of the Atlanta Campaign in May, 75 percent of Nathan B. Forrest's cavalry was armed with rifles, only a small percentage with breechloaders.

Major Gen. Nathan B. Forrest — Commanding Meridian, Mississippi May 25, 1864[27]

	1st Div.	3rd Div.	Holston Brig.	Escort Com.
.69 cal. Muskets	24	18		
.58 cal. Ashville Rifles	482	870	33	
.54 cal. Mississippi Rifles	1,475	854	265	
Shotguns		2		
.54 cal. Hall Carbines	4			
.52 cal. Sharps Carbines	725	400		63
.50 cal. Colt Carbines	11			
.56 cal. Burnside Carbines	52			5
.70 cal. Belgian Rifles	3	47		
.51 cal. Maynard Carbines	67			5
.52 cal. Hall Rifles		18		
French Pistols	190	163		
.44 cal. Revolvers	291	150	14	31
.36 cal. Revolvers	701	337	24	32
.54 cal. Holster Pistols	24			
Sabres		188	7	13

In September, two of the cavalry regiments in Forrest's Second Brigade, commanded by Colonel Robert McCulloch, were armed as follows:[28]

	18th Miss.	19th Miss.
.69 cal. Muskets	17	20
.54 cal. Austrian Rifles	72	136
.577 cal. Enfield Rifles	65	89
.54 Mississippi Rifles	5	18
Hall Carbines	1	
Sharps Carbines	91	57
Colt Carbines	2	
Burnside Carbines	19	11
Maynard Carbines	1	4
Union Carbines		19
Percussion Pistols	3	8
.44 cal. Revolvers	52	16
.36 cal. Revolvers	103	14
Sabres	6	30

In the Confederate field inspections report of May 25, 1864, about 25 percent of Nathan B. Forrest's cavalry was armed with captured Yankee carbines.
National Archives collection

REGIMENTAL INVENTORIES[29]

During the summer and fall of 1864, the Union cavalry and mounted infantry regiments were inventorying over 86,000 carbines in the following regiments:

1st U.S.	248 Sharps
3rd U.S.	154 Starr
	327 Sharps
4th U.S.	236 Spencer
	4 Burnside
1st U.S. Colored	49 Burnside
2nd U.S. Colored	588 Merrill
3rd U.S. Colored	105 Burnside
	28 Cosmopolitan
	540 Smith
4th U.S. Colored	60 Smith
	604 Burnside
1st Ark.	1,023 Starr
2nd Ark.	71 Sharps
	162 Smith
	129 Cosmopolitan
	82 Burnside
	417 Gallager
3rd Ark.	9 Sharps
	515 Starr
4th Ark.	3 Sharps
	10 Hall
1st Ala.	527 Smith
1st Calif.	657 Sharps
2nd Calif.	611 Sharps
1st Colo.	71 Sharps
2nd Colo.	86 Merrill
	805 Starr
3rd Colo.	58 Starr
1st Del.	169 Merrill
1st Dakota	141 Sharps
	3 Hall
2nd Ill.	237 Sharps
	633 Burnside
3rd Ill.	120 Sharps
	86 Burnside
4th Ill.	419 Sharps
5th Ill.	255 Sharps
	3 Cosmopolitan
6th Ill.	110 Sharps
	656 Spencer
7th Ill.	647 Sharps
9th Ill.	79 Sharps
	3 Hall
	259 Spencer
10th Ill.	465 Sharps
11th Ill.	24 Sharps
	819 Smith
	25 Merrill
12th Ill.	977 Spencer
13th Ill.	108 Gallager
14th Ill.	22 Sharps
	35 Burnside
15th Ill.	359 Sharps
	38 Burnside
16th Ill.	16 Sharps
1st Ind.	276 Sharps
	48 Burnside
2nd Ind.	14 Spencer
	59 Smith
	7 Merrill
3rd Ind.	192 Burnside
4th Ind.	213 Smith
5th Ind.	483 Sharps
6th Ind.	14 Sharps
	497 Burnside
	602 Gallager
	86 Maynard
7th Ind.	874 Merrill
8th Ind.	108 Spencer
	58 Joslyn
9th Ind.	146 Gallager
10th Ind.	711 Burnside
11th Ind.	100 Maynard
12th Ind.	273 Merrill
1st Iowa	1,468 Sharps
2nd Iowa	70 Sharps
	688 Spencer
3rd Iowa	643 Burnside
4th Iowa	545 Sharps
	644 Spencer
5th Iowa	383 Sharps
6th Iowa	63 Burnside
7th Iowa	486 Gallager
8th Iowa	120 Burnside
	461 Cosmopolitan
9th Iowa	592 Starr
2nd Kan.	275 Sharps
	86 Smith
	25 Cosmopolitan
	13 Merrill
5th Kan.	283 Sharps
	178 Starr
	2 Maynard
6th Kan.	252 Sharps
	170 Cosmopolitan
	49 Merrill
7th Kan.	121 Sharps
	387 Starr
	4 Maynard
	33 Musketoon
9th Kan.	49 Sharps
	70 Smith
	444 Gallager
	9 Pistol Carbine
10th Kan.	520 Smith
	143 Gallager
11th Kan.	88 Sharps
	78 Hall
	86 Starr
14th Kan.	153 Sharps
	124 Smith
	157 Cosmopolitan
	163 Gallager
15th Kan.	933 Sharps
	55 Hall
16th Kan.	70 Burnside
	95 Starr
	54 Smith
	310 Gallager
1st Ky.	2 Sharps
	5 Musketoon
2nd Ky.	158 Sharps
	50 Merrill
	52 Spencer
	58 Joslyn
3rd Ky.	389 Sharps
4th Ky.	10 Burnside
	143 Sharps
	17 Smith
5th Ky.	270 Burnside
	43 Smith
6th Ky.	24 Burnside
	166 Ballard
	71 Sharps
	51 Smith
7th Ky.	151 Burnside

The Year of Attrition, 1864

The 18th Pennsylvania Cavalry in winter camp, March 1864. By the summer campaign, the regiment had been issued Spencer carbines.
U.S. Military History Institute, MOLLUS

	24 Spencer	**Potomac Home Guard**			71 Hall
	87 Sharps		33 Sharps		66 Starr
11th Ky.	78 Spencer	**1st Miss.**	136 Sharps	**8th Mo.**	637 Cosmopolitan
	5 Burnside	**2nd Minn.**	935 Smith	**10th Mo.**	158 Gibbs
12th Ky.	5 Ballard		170 Sharps	**11th Mo.**	213 Sharps
13th Ky.	762 Ballard	**Minn. Ind. Cav.**			46 Starr
16th Ky.	119 Ballard		255 Smith		332 Merrill
4th Ky. Mtd. Inf.			79 Burnside	**12th Mo.**	719 Starr
	74 Merrill		53 Sharps	**13th Mo.**	1,030 Starr
	200 Ballard	**Brackett Batt'l**		**14th Mo.**	85 Starr
	32 Spencer		174 Burnside		165 Smith
30th Ky. Mtd. Inf.		**8th Minn. Mt. Inf.**		**1st Mo. State Militia**	
	39 Smith		71 Smith		118 Burnside
	387 Ballard	**1st Mich.**	271 Spencer		56 Musketoon
37th Ky. Mtd. Inf		**2nd Mich.**	402 Spencer	**2nd Mo. State Militia**	
	527 Ballard	**3rd Mich.**	7 Musketoon		88 Sharps
39th Ky. Mtd. Inf.			888 Starr		244 Hall
	49 Musketoon		320 Merrill	**3rd Mo. State Militia**	
40th Ky. Mtd. Inf.		**4th Mich.**	41 Sharps		500 Wesson
	49 Cosmopolitan		230 Spencer	**4th Mo. State Militia**	
	40 Burnside	**5th Mich.**	163 Spencer		40 Musketoon
45th Ky. Mtd. Inf.		**6th Mich.**	248 Spencer		24 Pistol carbine
	275 Ballard	**7th Mich.**	76 Spencer	**6th Mo. State Militia**	
52nd Ky. Mtd. Inf.		**8th Mich.**	7 Musketoon		263 Hall
	710 Ballard		21 Burnside		132 Wesson
54th Ky. Mtd. Inf.		**9th Mich.**	55 Burnside		70 Burnside
	100 Ballard		246 Spencer		134 Starr
1st La.	113 Burnside	**10th Mich.**	13 Smith	**7th Mo. State Militia**	
	259 Sharps		25 Gallager		17 Sharps
2nd La.	235 Musketoon		24 Burnside		631 Smith
1st Maine	82 Burnside	**11th Mich.**	369 Spencer	**8th Mo. State Militia**	
1st Mass.	6 Burnside		357 Starr		60 Wesson
2nd Mass.	20 Burnside	**1st Mo.**	169 Sharps	**9th Mo. State Militia**	
	318 Sharps	**2nd Mo.**	197 Sharps		302 Merrill
3rd Mass.	276 Sharps		414 Starr	**32th Mo. Mtd. Inf.**	
4th Mass.	171 Sharps	**3rd Mo.**	713 Sharps		35 Sharps
31st Mass. Mtd. Inf.		**4th Mo.**	355 Sharps	**1st Neb.**	3 Sharps
	2 Burnside		45 Gallager		50 Merrill
40th Mass. Mtd. Inf.		**6th Mo.**	148 Sharps	**1st N.H.**	95 Sharps
	18 Sharps		130 Burnside		5 Burnside
1st Md.	40 Sharps	**7th Mo.**	267 Sharps		

A detachment of the 1st Regular Cavalry at Brandy Station, Virginia, in February 1864. At this time, the unit was issued Sharps carbines.
U.S. Military History Institute, MOLLUS

Unit	Count & Type	Unit	Count & Type	Unit	Count & Type
1st N.M.	6 Sharps	20th N.Y.	674 Starr	1st Oregon	77 Sharps
1st N.J.	52 Sharps		60 Sharps	3rd Penn.	304 Sharps
	294 Burnside	21st N.Y.	182 Burnside		11 Burnside
	13 Starr		15 Smith	5th Penn.	10 Merrill
2nd N.J.	313 Spencer	22nd N.Y.	48 Sharps		204 Sharps
3rd N.J.	6 Burnside		10 Burnside	7th Penn.	392 Spencer
	61 Spencer	23rd N.Y.	61 Starr		27 Burnside
1st N.Y. Veteran		24th N.Y.	238 Starr		77 Sharps
	40 Sharps	25th N.Y.	226 Burnside	8th Penn.	31 Burnside
	132 Smith	1st Nev.	22 Starr	9th Penn.	7 Sharps
	5 Merrill		290 Sharps		106 Burnside
	84 Starr	1st Ohio	8 Burnside		541 Starr
	68 Burnside		423 Sharps	11th Penn.	80 Merrill
1st Lincoln N.Y.		2nd Ohio	241 Spencer		36 Sharps
	3 Smith		72 Burnside		49 Burnside
2nd N.Y.	188 Spencer	3rd Ohio	176 Spencer		78 Spencer
2nd N.Y. Vet.			50 Burnside	12th Penn.	886 Burnside
	443 Sharps		26 Sharps	13th Penn.	25 Gallager
3rd N.Y.	147 Sharps	4th Ohio	21 Spencer		94 Burnside
	59 Burnside		253 Burnside	14th Penn.	596 Burnside
5th N.Y.	218 Spencer		180 Sharps		12 Spencer
7th N.Y.	600 Sharps	5th Ohio	433 Burnside	15th Penn.	363 Sharps
9th N.Y.	9 Starr		77 Sharps		95 Burnside
	? Sharps & Hankins	6th Ohio	105 Burnside	16th Penn.	12 Burnside
			3 Smith		7 Starr
10th N.Y.	15 Burnside	7th Ohio	275 Burnside	17th Penn.	9 Starr
	47 Smith		2 Ballard		50 Spencer
11th N.Y.	976 Burnside	8th Ohio	2 Burnside	18th Penn.	31 Burnside
	70 Sharps & Hankins		101 Spencer		325 Spencer
			478 Cosmopolitan	19th Penn.	25 Sharps
12th N.Y.	473 Burnside	9th Ohio	515 Smith		43 Starr
	127 Starr	10th Ohio	586 Sharps	20th Penn.	52 Burnside
13th N.Y.	516 Sharps	11th Ohio	264 Spencer	21st Penn.	68 Gallager
	4 Spencer		108 Smith	22nd Penn.	354 Sharps
14th N.Y.	3 Sharps		90 Merrill		424 Smith
	207 Burnside		18 Musketoon		20 Starr
15th N.Y.	344 Burnside	12th Ohio	93 Ballard		98 Burnside
	48 Starr		228 Spencer	1st R.I.	137 Sharps
16th N.Y.	688 Sharps	McLaughlin Squadron		1st Tenn.	416 Burnside
18th N.Y.	174 Burnside		88 Ballard		45 Sharps
19th N.Y.	185 Spencer		18 Burnside		5 Gallager

The Year of Attrition, 1864

Members of Company C, 3rd Pennsylvania Cavalry at Brandy Station in March 1864. In addition to their sabers, the regiment was armed with Sharps carbines and Colt revolvers.
U.S. Military History Institute, MOLLUS

Regiment	Qty	Type	Regiment	Qty	Type	Regiment	Qty	Type
2nd Tenn.	144	Burnside	**13th Tenn.**	381	Joslyn	**5th W.V.**	32	Smith
	3	Sharps	**1st Tenn. Mtd. Inf.**			**7th W.V.**	41	Lindner
	213	Gallager		104	Merrill		87	Burnside
	2	Merrill	**2nd Tenn. Mtd. Inf.**			**1st Vt.**	16	Spencer
3rd Tenn.	403	Sharps		134	Merrill	**1st Wisc.**	22	Maynard
4th Tenn.	408	Gallager	**4th Tenn. Mtd. Inf.**				12	Merrill
5th Tenn.	87	Smith		60	Gallager		200	Warner
	93	Merrill	**1st Texas**	508	Burnside		4	Smith
6th Tenn.	394	Sharps	**1st W.V.**	36	Sharps		142	Joslyn
7th Tenn.	7	Sharps		3	Smith		392	Burnside
	3	Cosmopolitan		357	Spencer		1	Spencer
	15	Gallager	**2nd W.V.**	4	Sharps	**2nd Wisc.**	603	Sharps
8th Tenn.	745	Burnside		73	Smith	**3rd Wisc.**	40	Musketoon
9th Tenn.	291	Gallager		89	Joslyn		358	Cosmopolitan
10th Tenn.	78	Burnside		460	Spencer		36	Gallager
	261	Merrill	**3rd W.V.**	184	Sharps		265	Merrill
	140	Maynard		75	Smith		623	Burnside
12th Tenn.	591	Sharps		81	Gallager	**4th Wisc.**	422	Burnside
	25	Merrill		383	Spencer			

Footnotes — The Year of Attrition, 1864

[1] NARG 156-103 and ORs Series III, Vol. 4, p. 1167
[2] NARG 156-215
[3] NARG 108-75
[4] Ibid.
[5] Ibid.
[6] Ibid.
[7] NARG 156-215
[8] NARG 108-75
[9] Ibid.
[10] Ibid.
[11] NARG 156-215
[12] NARG 156-21, Box 254
[13] ORs Vol. 36, Series I, Part I, p. 879
[14] Ibid., pp. 186-187 and 1096, NARG 156-110, June 1864
[15] ORs Vol. 38, Series I, Part I, pp. 89-117
[16] Stephen Z. Starr, *The Union Cavalry in the Civil War, Vol. III*, Baton Rouge, LA, 1985, p. 478
[17] NARG 156-110, Quarterly Reports of June 30 and Sept 30, 1864
[18] NARG 156-21
[19] Stephen Z. Starr, *The Union Cavalry in the Civil War, Vol. II*, Baton Rouge, LA, 1981, p. 252
[20] NARG 156-21
[21] Ibid., Box 242
[22] NARG 156-110
[23] ORs Vol 43, Section I, Part I, pp. 529-530 and Stephen Z. Starr. Vol. II, op.cit., pp. 260-261
[24] ORs Vol. 43, Series I, Part I, p. 60
[25] ORs Vol. 41, Series I, Part I, pp. 939-942
[26] NARG 109- M935 (18 rolls)
[27] Ibid.
[28] Ibid.
[29] NARG 156-110 reports of June and September 1864 plus NARG 108-77

Judson Kilpatrick led Sherman's cavalry during the 1865 campaign.
U.S. Military History Institute, MOLLUS

THE END OF HOSTILITIES
◆ 1865 ◆

By March, the Union cavalry was rapidly preparing for the upcoming spring campaign. The campaign would see the surrender of all Confederate forces within the next four months. The following charts reflect the armament of the Union cavalry at the military department level as of March 1, 1865.

Western Military Departments[1]
March 1, 1865

Carbine	Cumberland	Missouri	Mississippi	Military Division of Mississippi	Arkansas	Northwest
Sharps	—	812	3,930	1,710	2,266	471
Colt .44 cal.	—	37	—	—	479	—
Hall	—	151	—	—	10	180
Merrill	163	123	—	100	50	—
Joslyn	—	118	—	6	—	—
Starr	—	851	174	378	2,057	—
Gallager	—	278	114	74	362	—
Burnside	78	262	696	1,356	—	372
Cosmopolitan	—	560	123	—	580	—
Wesson	—	127	—	—	—	—
Smith	231	939	797	154	56	1,319
Spencer	—	739	896	8,629	764	—
Musketoon	—	50	—	—	—	—
Ballard	75	—	—	76	—	—
Maynard	—	—	—	1,651	800	—
Warner	—	—	—	361	—	—

By February 1865, over 2,400 of the 2nd Model Maynard carbines were in field service. A month later, they were issued to the 2nd California Cavalry to replace their old Model 1853 Sharps. The Maynards were generally looked upon as easy to load and very accurate. *U.S. Military Academy collection*

Eastern Military Departments[2]
March 1, 1865

Carbines	Potomac	Middle Military Division	Virginia and North Carolina	Washington	South	Ohio
Sharps	3,237	3,913	1,588	1,673	200	52
Spencer	1,508	5,820	844	390	—	—
Burnside	437	3,050	—	215	—	—
Starr	—	31	571	45	—	—
Smith	35	173	—	—	—	—
Cosmopolitan	—	10	—	—	—	—
Gallager	—	749	—	—	—	—
Ballard	—	—	283	—	—	163

Total Carbines at Department Level
March 1, 1865

Ballard	597	Musketoon	50
Burnside	6,456	Merrill	440
Colt	516	Sharps	19,852
Cosmopolitan	1,273	Smith	3,686
Gallager	1,577	Spencer	19,590
Hall	341	Starr	4,107
Joslyn	124	Warner	361
Maynard	2,451	Wesson	127

NOTE: Kilpatrick's cavalry was attached to Sherman's army and his carbines are not reflected in these numbers. His cavalry, as of the first of March, was armed with Sharps, Spencer, Joslyn, Smith and Ballard carbines. Two hundred fifty seven (257) of the Sharps carbines reported in the Department of Mississippi were actually Sharps & Hankins carbines in the 11th New York Cavalry.

Appomattox Campaign

On the 29th of March, Phil Sheridan left the Petersburg lines with three divisions of cavalry. Sheridan's goal was to destroy both the South Side and Danville Railroads. His divisional commanders were Thomas Devin, George Crook and George Custer. In addition to Sheridan's 9,000 cavalrymen, Ranald Mackenzie's cavalry division of the Army of Virginia was attached to Sheridan on April 1. When the cavalry departed from Hancock Station, each cavalryman had been issued five days' rations and forty rounds of ammunition.[3]

As the charts on the following page show, sixty percent of Sheridan's cavalry in the Appomattox Campaign were totally or partially armed with Spencer carbines. The Spencer's rate of fire would play a critical role in the upcoming campaign.

Near Dinwiddie Court House on the last day of March, the Brigades of Smith and Davies were attacked and driven back by Confederate forces led by George Pickett. To help cover their withdrawal, Sheridan used Gregg's and Gibbs' brigades armed with Sharps carbines to check Pickett's attack. When the main attack on Sheridan's position came late in the day, the brigades of Capehart, Pennington, Gibbs, Gregg and Smith were able to break up the Confederate attacks with concentrated fire from their Sharps and Spencer carbines. While the Union cavalry was hard pressed in this engagement, their carbine fire made the difference between defeat and

Phil Sheridan's Cavalry[4]
March 29 - April 9, 1865

1st Division — Thomas Devin
Brigades

Peter Stagg
1st Mich. Spencer
5th Mich. Spencer
6th Mich. Spencer
7th Mich. Spencer

Charles Fitzhugh
6th N.Y. Sharps
9th N.Y. Sharps
19th N.Y. Spencer
17th Penn. Sharps/Spencer
20th Penn. Burnside

Alfred Gibbs
2nd Mass. Spencer
6th Penn. Sharps
1st U.S. Sharps
5th U.S. Sharps
6th U.S. Sharps

2nd Division — George Crook
Brigades

Henry Davies, Jr.
1st Penn. Sharps
1st N.J. Spencer
10th N.Y. Sharps/Spencer
24th N.Y. Sharps/Burnside

J. Irvin Gregg
4th Penn. Spencer
8th Penn. Sharps
16th Penn. Spencer
21st Penn. Sharps

Charles Smith
1st Maine Spencer
2nd N.Y. Mtd. Rifles Sharps
6th Ohio Sharps/Burnside
13th Ohio Sharps

3rd Division — George Custer
Brigades

Alexander Pennington
1st Conn. Spencer
3rd N.J. Spencer
2nd N.Y. Spencer
2nd Ohio Spencer

William Wells
8th N.Y. Sharps/Burnside
15th N.Y. Sharps
1st Vt. Sharps/Spencer

Henry Capehart
1st N.Y. (Lincoln) Sharps/Spencer
1st W.V. Spencer
2nd W.V. Spencer
3rd W.V. Spencer

Ranald Mackenzie — Army of Virginia — Cavalry
Brigades

Robert West
20th N.Y. Starr
5th Penn. Sharps/Spencer

Samuel Spear
1st Md. Sharps
11th Penn. Spencer
1st D.C. Henry/Spencer

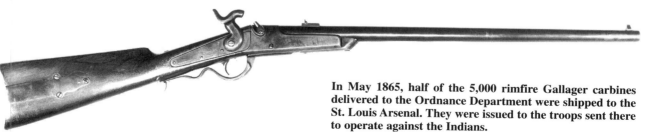

In May 1865, half of the 5,000 rimfire Gallager carbines delivered to the Ordnance Department were shipped to the St. Louis Arsenal. They were issued to the troops sent there to operate against the Indians.

U.S. Military Academy collection

a draw. During the early hours of the next day, Pickett withdrew a short distance to Five Forks. Here, Pickett established his position behind breastworks and awaited Sheridan's attack.

Sheridan's plan of action at Five Forks first called for the infantry of the Fifth Army Corps to make the main assault with diversionary charges by the Union cavalry. The dismounted cavalry attacks were carried out by Stagg, Fitzhugh, Gibbs and Pennington's brigades. Pennington's men, totally armed with Spencer carbines, nearly ran out of ammunition. A supply of Spencer cartridges was received and distributed to his men in the line of battle while heavily engaged.[5] A mounted charge was conducted by Wells and Capehart of Custer's division. All this action took place as the infantry was attacking on the right of the cavalry.

These combined attacks broke through Pickett's breastworks and netted Sheridan nearly 6,000 prisoners. The Union casualties were listed at about 1,000 with a third coming from the cavalry. In his report of the Appomattox Campaign, Brevet Major General Wesley Merritt stated that, during this campaign, Fitzhugh's men mounted the Confederate breastworks in the face of enemy fire, tearing down their colors and capturing two pieces of artillery and nearly 1,000 prisoners.[6] The 1st Connecticut Cavalry, under Pennington's command, captured two 3-inch rifled cannons. These two cannon were captured by Major Goodwin and Lieutenant Lanfare who were both awarded the Medal of Honor for their deeds.[7]

Five days later, on April 6, the Union cavalry and infantry made a major contribution towards Lee's upcoming surrender to Grant at Appomattox Court House. The Union cavalry and infantry surrounded and captured a large portion of Lee's army at Sayler's Creek. In this engagement, orders read that every cavalryman with a Spencer carbine was to be issued 125 rounds of ammunition.[8]

Custer states, in his official report of the battle at Sayler's Creek, that his division captured seven general officers, including Lieutenant General Richard Ewell, 15 pieces of artillery and 31 battle flags. Colonel Pennington's brigade captured 190 commissioned officers and 1,834 enlisted men as well as capturing or destroying about 300 wagons.[9] The capture of Ewell and his staff was credited to Captain Samuel Stevens, Company C, 1st New York (Lincoln) Cavalry. Captain Edwin F. Savacool, Company K of the same regiment, was credited with the

Wesley Merritt was second-in-command of the Cavalry Corps of the Army of the Potomac in the Appomattox Campaign. *U.S. Military History Institute, MOLLUS*

capture of three Confederate battle flags. Savacool was awarded the Congressional Medal of Honor.[10] On the tenth of April, the day after Lee's surrender, the 1st Delaware Cavalry, guarding the Baltimore and Ohio Railroad, requested an issue of Spencer carbines for possible guerrilla action. As late as the end of April, Devin's and Custer's troopers, now stationed at City Point, were issued 225 Spencers and 184 Sharps carbines as well as 1,926 horses. With the war at an end in Virginia, the Army of the Potomac notified Washington that no more carbines were needed in the field. This action occurred on May 5 and, on the following day, carbine ammunition for the Burnsides, Merrills and Starrs was sent to the northern depots and arsenals.[11]

Wilson's Selma Campaign

Late in 1864, James H. Wilson transferred from the East to take command of Sherman's cavalry. As of the first of February 1865, the four cavalry divisions that would play a role in the upcoming spring Selma Campaign were as follows:

James H. Wilson, Commanding[12]
As of February 1, 1865

1st Division — Edward McCook

1st Brigade	2nd Brigade	Brigade Armament	
8th Iowa	2nd Indiana	Spencer .. 1,189	Smith....... 145
2nd Michigan	4th Indiana	Sharps 614	Ballard Rifle .. 86
6th Kentucky	4th & 7th Kentucky	Burnside.. 1,009	Colt........ 149
4th Kentucky Mounted Inf.	1st Wisconsin	Maynard 4	Rifle Muskets . 89
		Warner..... 343	

2nd Division — Eli Long

1st Brigade	2nd Brigade	Brigade Armament
98th Illinois Mtd. Inf.	4th Michigan	Spencer 3,585
123rd Illinois Mtd. Inf.	3rd Ohio	Sharps 1,113
17th Indiana Mtd. Inf.	4th Ohio	Spencer Rifle.. 1,387
72nd Indiana Mtd. Inf.	7th Pennsylvania	

4th Division — Emory Upton

1st Brigade	2nd Brigade	Brigade Armament
3rd Iowa	5th Iowa	Spencer 417
4th Iowa	1st Ohio	Sharps 83
10th Missouri	7th Ohio	Burnside....... 394

5th Division — Edward Hatch

1st Brigade	2nd Brigade	Brigade Armament	
3rd Illinois	6th, 7th, 9th Ill.	Spencer... 1,461	Maynard ... 775
11th Indiana	2nd Iowa	Sharps 772	Starr 655
12th Missouri	12th Tennessee	Burnside 98	Colt Rifles .. 202

Major Chambliss, Special Inspector of Cavalry for the Military Division of Mississippi, from his office in Louisville, Kentucky, wrote Wilson on January 27 stating that he had on hand between 500 and 600 Spencer carbines for issue. In addition, he wrote that the department was entitled to a portion of the 3,000 Spencers per month that were being received from the factory.[13] As of the first of February, Wilson's cavalry was reporting 6,721 Spencers and, a month later, this total had increased nearly by 2,000, to 8,629.[14] Wilson, from Gravelly Springs, Alabama, on February 17, requested that his cavalrymen also be equipped with the Blakeslee cartridge boxes for the Spencer carbines.[15]

Wilson's plans called for taking three divisions of cavalry on his Selma raid. Since the 5th division was not to take part in the raid, Colonel Datus Coon of the Second Brigade of this division offered to turn over all 1,400-plus Spencer carbines of his brigade for use by the raiders. Colonel Coon's offer of March 1 was given to John Croxton, temporarily in command of McCook's First Division. A week later, Brigadier General Edward Hatch, Colonel Coon's division commander, gave his consent to turn over his Spencers. In his reply, Hatch states: "If an order can be obtained to send the command to the rear to remount, as much as I should dislike to lose the only good arm this division has, I should consent, on the receipt of that order, to turn the arms over for the good of the service."[16] Wilson greatly appreciated

Hatch's official offer. When Wilson left on his raid on March 22 with the 1st, 2nd and 4th Divisions, they were mostly armed with Spencers. They did, however, have about 700 Sharps and 300 Burnsides besides their Spencer carbines and rifles. The Spencer rifles were in Wilder's old mounted infantry brigade. After turning over its Spencers, Hatch's division was ordered to be remounted and receive new arms from Major Chambliss at Louisville.

On the afternoon of April 2, Wilson's cavalry arrived before the defenses of Selma, which were commanded by Lieutenant General Nathan B. Forrest. At 5:00 p.m., Eli Long, commanding the 2nd division, launched his dismounted attack against the main defense of breastworks. Leading the advance, Long had 1,550 officers and men of the 17th Indiana Mounted Infantry, 98th and 123rd Illinois Mounted Infantry and the 4th Ohio and 7th Pennsylvania Cavalry. Long's cavalrymen had to cross 600 yards of open ground before reaching the Confederate defenses, but within 25 minutes the enemy's breastworks had been captured. One hundred and fifty yards from the Confederate defenses, Long was wounded and carried from the field. In his official report, Long states: "We moved forward steadily until within a short range when a rapid fire was opened by our Spencers and with a cheer the men started for the works on a run, sweeping forward in solid line over fences and ravine, scaling the stockade and on the works with resistless force." While the 4th Ohio and 7th Pennsylvania were armed with Spencer carbines, the Spencer rifles were in the hands of Colonel Abraham O. Miller's mounted infantry regiments.[17] The taking of the Confederate breastworks had cost Long's division 39 killed and 261 wounded.

With Long's attack in progress, Emory Upton sent forward the dismounted 3rd Iowa and the 10th Missouri. As soon as the enemy position in their front was captured, the 4th Iowa charged into the city capturing several pieces of artillery and several hundred prisoners. Upton's casualties came to 8

James Harrison Wilson led the Union cavalry in the Selma Raid in March 1865.
U.S. Military History Institute, MOLLUS

killed, 65 wounded and 9 missing. Leading Upton's troops were Wilson and his escort, the 4th U.S. Cavalry, which charged with drawn sabres right through the gap in the Confederate line. The capture of Selma netted Wilson 150 officers and 2,700 enlisted personnel. Forrest escaped capture. Wilson's raid continued until he reached Macon, Georgia, on April 21 where, under a flag of truce, he learned of the armistice between Generals Sherman and Johnston.[18] The war was over.

Model 1860 Spencer carbine sn. 54,589 was issued to B.O. Mitchell, Company A, 3rd Kentucky Cavalry in March 1865. The 3rd was part of Kilpatrick's Cavalry in Sherman's Carolina Campaign. *Author's collection*

REGIMENTAL INVENTORIES — 1865[19]

As the military department totals reflected from March 1, the three main carbines in the cavalry were the Sharps, Spencer and Burnside. The following totals were taken from reports received between January and May 1865.

Sharps Carbines

Regiment	Count
2nd Arkansas	666
2nd California	131
3rd Illinois	35
4th Illinois	394
5th Illinois	509
7th Illinois	603
9th Illinois	88
10th Illinois	907
17th Illinois	649
1st Indiana	139
3rd Indiana	40
4th Indiana	272
5th Indiana	254
8th Indiana	34
1st Iowa	672
9th Iowa	721
2nd Kentucky	14
7th Kentucky	154
5th Kansas	94
9th Kansas	73
11th Kansas	81
15th Kansas	821
1st Louisiana	532
1st Maryland	408
2nd Maryland	580
3rd Maryland	59
1st Massachusetts	217
2nd Massachusetts	24
3rd Massachusetts	507
4th Massachusetts	30
2nd Minnesota	179
1st Mississippi	259
1st Missouri	598
1st Missouri State	21
3rd Missouri State	75
8th Missouri State	12
8th Michigan	400
1st N.H.	122
2nd N.J.	412
4th N.Y.	135
6th N.Y.	488
8th N.Y.	217
16th Missouri	16
Bracketts Battalion	168
Adams Independent	126
13th Ohio	425
Upper Potomac Force	494
1st Battalion Nevada	159
15th N.Y.	601
16th N.Y.	530
22nd N.Y.	352
24th N.Y.	105
3rd Ohio	142
4th Ohio	140
6th Ohio	155
10th Ohio	250
1st Pennsylvania	209
2nd Pennsylvania	491
3rd Pennsylvania	408
5th Pennsylvania	348
6th Pennsylvania	650
8th Pennsylvania	316
13th Pennsylvania	267
15th Pennsylvania	381
17th Pennsylvania	467
19th Pennsylvania	317
21st Pennsylvania	480
1st Rhode Island	143
1st Tennessee	286
6th Tennessee	675
10th Tennessee	176
12th Tennessee	20
1st Texas	255
2nd U.S.	234
3rd U.S.	209
5th U.S.	233
6th U.S.	215
3rd U.S. Colored	746
4th U.S. Colored	59
5th U.S. Colored	152
1st Vermont	397
1st W. Virginia	34
3rd W. Virginia	10
2nd Wisconsin	23

Continued on next page

9th N.Y.	512
10th N.Y.	160
1st N.Y. Lincoln	179
1st N.Y. Mounted Rifle	633
2nd N.Y. Mounted Rifle	366
2nd N.Y. Vet.	716
15th Missouri	32

Burnside Carbines

2nd Illinois	578	1st N.Y. (Lincoln)	9
3rd Illinois	12	5th Missouri	188
6th Illinois	420	1st Dakota	139
9th Illinois	333	Bracketts Battalion	160
10th Illinois	15	Adams Independent	189
11th Illinois	160	14th N.Y.	176
14th Illinois	344	21st N.Y.	468
16th Illinois	153	24th N.Y.	219
17th Illinois	268	25th N.Y.	322
10th Indiana	372	6th Ohio	98
13th Indiana	776	7th Ohio	250
16th Indiana Mtd. Inf.	479	8th Ohio	622
6th Iowa	63	McLaughton Squadron	50
7th Iowa	38	3rd Pennsylvania	186
3rd Kentucky	38	9th Pennsylvania	23
6th Kentucky	12	12th Pennsylvania	685
7th Kentucky	85	13th Pennsylvania	57
39th Kentucky Mtd. Inf.	57	14th Pennsylvania	510
16th Kansas	70	20th Pennsylvania	632
3rd Maryland	708	3rd Rhode Island	456
2nd Maine	529	1st Tennessee	506
14th Missouri	345	2nd Tennessee	321
1st Missouri State	111	7th Tennessee	172
6th Missouri State	15	1st Texas	533
9th Michigan	79	3rd U.S. Colored	92
8th N.Y.	44	4th U.S. Colored	704
11th N.Y.	190	2nd Wisconsin	18
		4th Wisconsin	703

Spencer Carbines

1st Connecticut	404	2nd N.Y.	613
1st Florida	52	5th N.Y.	263
2nd Illinois	76	10th N.Y.	400
3rd Illinois	6	1st N.Y. Vet.	892
6th Illinois	57	1st N.Y. (Lincoln)	22
12th Illinois	598	18th N.Y.	463
17th Illinois	80	19th N.Y.	435
2nd Indiana	166	21st N.Y.	13
4th Indiana	8	22nd N.Y.	45

Regiment	Count	Regiment	Count
7th Indiana	396	25th N.Y.	7
8th Indiana	150	1st Ohio	454
9th Indiana	15	2nd Ohio	356
2nd Iowa	621	3rd Ohio	384
3rd Iowa	761	4th Ohio	442
4th Iowa	676	7th Ohio	104
5th Iowa	544	8th Ohio	30
8th Iowa	563	9th Ohio	591
2nd Kentucky	66	10th Ohio	16
3rd Kentucky	178	11th Ohio	77
4th Kentucky	222	12th Ohio	345
6th Kentucky	377	4th Pennsylvania	425
4th Kentucky Mtd. Inf.	440	5th Pennsylvania	197
7th Kansas	441	7th Pennsylvania	766
1st Maine	243	9th Pennsylvania	47
2nd Maine	122	11th Pennsylvania	500
2nd Massachusetts	423	14th Pennsylvania	38
4th Massachusetts	452	16th Pennsylvania	376
10th Missouri	455	17th Pennsylvania	30
9th Missouri State	13	18th Pennsylvania	335
1st Michigan	352	1st Rhode Island	9
2nd Michigan	386	2nd Tennessee	15
3rd Michigan	760	12th Tennessee	4
4th Michigan	486	1st Vermont	121
5th Michigan	234	1st W. Virginia	329
6th Michigan	309	2nd W. Virginia	398
7th Michigan	268	3rd W. Virginia	481
1st New Jersey	475	6th W. Virginia	68
2nd New Jersey	310	7th W. Virginia	360
3rd New Jersey	585	1st Wisconsin	282
1st N.Y. Mtd. Rifles	276	2nd Wisconsin	650

Blakeslee Cartridge Boxes

In December of 1864, the late Colonel of the 1st Connecticut Cavalry, Erastus Blakeslee, received a patent for a cartridge box that contained from six to thirteen tubes. Each tube contained seven Spencer cartridges for quick loading of the Spencer. Most of the Blakeslee cartridge boxes for the cavalry contained ten tubes or 70 rounds of Spencer ammunition.

These Blakeslee cartridge boxes were in great demand for the spring campaign. The Ordnance Department, in January, was requested to send to Memphis, Tennessee, for issue to the 2nd New Jersey Cavalry, 900 Blakeslee cartridge boxes and 800 Spencer carbines. The following month, the Memphis depot requested 3,000 more Blakeslees plus 1,200 Spencer carbines. They received 1,500 Blakeslee boxes and the carbines that were sent to Wilson for his Selma raid. Some of the ten-tube Blakeslee cartridge boxes were issued to Captain George Nobles, Company D, 2nd Wisconsin Cavalry.[20]

Spencer

Many cavalry officers were trying to obtain Spencer carbines for their commands. The following are two examples of their attempts. On February 1, 1865, Colonel John McConnell of the 6th Illinois Cavalry, writing from 1st Brigade Headquarters at Memphis, Tennessee, requested that the War Department have his regiment replace their current inventory of 625 Sharps carbines for a like quantity of Spencer carbines. The answer came

Advertisement for the Blakeslee's cartridge boxes. The Wilson cavalry had a large number of these boxes during the Selma Raid.
National Archives collection

back that since the 6th was not one of the best cavalry regiments in point of discipline and efficiency in the Department of the Mississippi, its request for Spencers was denied. The colonel's Spencers had been turned in and were reissued to the 1st Division of Wilson's cavalry that went on the Selma raid. As of March, the 6th Illinois were reporting 57 Spencers and 420 Burnsides.[21] It appears that the 6th had not turned in all of their Spencers for Wilson's raid.

Near Savannah, Georgia, on January 3, Judson Kilpatrick wrote "no arms of any kind are to be had at this point as I expected there would be. The Joslyn carbine, with which the 9th Pennsylvania is armed, and the majority of my Sharps carbines, are utterly worthless. I earnestly request that 300 Spencer carbines to be sent to this point."[22] The only Spencer carbines in Kilpatrick's command at this time were with the 8th Indiana and 9th Michigan, while the 9th and 92nd Illinois Mounted Infantry were armed with Spencer Rifles. The balance of Kilpatrick's cavalry was armed with Sharps, Burnside, Smith and Joslyn carbines.

Kilpatrick left for the Carolina campaign without receiving any additional small arms. On March 12, Colonel Jordan, commanding the First Brigade, wrote his wife from Fayetteville, North Carolina that his whole brigade was about to be rearmed with Spencer carbines. On about the 28th of March, Kilpatrick received 1,000 Spencer carbines at Goldsboro with a promise of 1,300 more.[23] These Spencers were issued to the 2nd and 3rd Kentucky as well as the 9th Pennsylvania cavalry. The Spencer carbines came too late for Colonel William Hamilton of the 9th Ohio Cavalry whose regiment was armed with Smith carbines throughout the campaign. In his report of March 30, Colonel Hamilton wrote: "During the latter part of the campaign my command was rendered to a considerable extent ineffective on account of the lack of ammunition for our carbines [Smiths] a large portion of it having been rendered worthless by the rains which fell during the march. I regard the weapon for that reason, and for its liability to get out of repair, as one which should not be used in the service."[24] The 9th's Smith carbines were replaced in April with about 600 Spencer carbines.

With the war at an end, many of the cavalry regiments were mustered out of federal service while others were sent out West to help keep peace with the Indians. The following cavalry regiments turned in the following quantities of carbines:

Carbines Turned In — 1865[25]

3rd Colorado		7th Illinois		6th Kansas		14th Kansas	
Sharps	17	Burnside	33	Merrill	47	Spencer	22
Starr	169	Sharps	603	Cosmopolitan	142	Cosmopolitan	1
3rd Illinois		**9th Illinois**		Sharps	17	**14th Pennsylvania**	
Burnside	15	Gallager	103	**9th Kansas**		Starr (rimfire)	163
Sharps	32	**11th Indiana**		Gallager	169	Burnside	14
6th Illinois		Gallager	68	Sharps	10	Spencer	1
Burnside	733	Maynard	442	Smith	44	**12th Tennessee**	
Spencer	77	**2nd Iowa**		Maynard	1	Maynard	80
		Ballard	22			Sharps	180
		Spencer	40				

Ballard

When the 3rd Battalion of the 13th New York Heavy Artillery was formed in the fall of 1863, it was understood that these men would consist of seamen for marine artillery and Army gunboat service. As of 1865, Companies I, K, L and M of the 13th were attached to the Army's Naval Brigade, Army of the James. They were issued naval uniforms and armed with .44 caliber Ballard carbines and, probably, Ames model 1860 naval cutlasses. Four hundred of these cutlasses had been previously purchased for the gunboats. In March 1865, the

NCOs of the 13th New York Cavalry in 1865, at which time the regiment was armed with Spencer carbines.
U.S. Military History Institute, MOLLUS

13th New York Heavy Artillery was reporting 283 Ballard carbines on hand. The 13th operated aboard the steamers *Reno, Parks, Foster* and the *Burnside*. These steamers operated against guerilla activity along the James River and the Atlantic coastline.[26]

From the ordnance depot at Louisville, Kentucky on June 21, First Lieutenant Babbitt wrote Dyer requesting instructions on several hundred Ballard and Wesson carbines that had been turned in by Kentucky troops. Since the carbines were the property of the State of Kentucky, Lieutenant Babbitt requested permission to turn them over to the state and take receipt for them. This request was granted. Two years later, on November 30, 1867, the State of Kentucky had in inventory the following carbines: 320 Wessons, 1,674 Ballards .44 cal., 570 Ballards .56 cal., 2,505 Ballards .46 cal., 2,405 Ballard Musketoons .46 cal. and 2,000 Triplett & Scott with 22-inch barrels, as well as 22 various makes of carbines.[27]

Rimfire Gallager

Four of Edward Hatch's regiments that had not been sent on Wilson's raid were ordered to St. Louis. These 2,500 cavalrymen consisted of the 3rd Illinois, 11th Indiana, 12th Missouri and the 12th Tennessee. They arrived in the St. Louis area in early May armed with percussion Starr and Gallager carbines and Enfield rifles.

Needing better small arms to operate on the plains, Major Callender, commanding officer at the St. Louis Arsenal, placed a request on May 16 with the Ordnance Department for 2,500 Spencer carbines. Instead of the Spencer, the Ordnance Department directed Frankford Arsenal to send 2,500 rimfire Gallager carbines and a half-million .56-.50 rimfire cartridges from New York. By the end of May, the 12th Missouri and the 12th Tennessee had been issued their new Gallager carbines, and the 11th Indiana and 3rd Illinois received theirs by the fifth of June. Colonel Carnahan's 3rd Illinois was issued 630 Gallager carbines, New Model; 630 carbine cartridge boxes and 13,104 Spencer cartridges.[28]

The 11th Indiana and 12th Tennessee spent most of the summer in Missouri and near Fort Riley, Kansas, while the 3rd Illinois operated against hostiles in Minnesota and the Dakota Territory until October. By the fall, these three regiments were mustered out of federal service. Only the 12th Missouri saw action against the Indians. On the Powder River between September 1 and 5, they clashed several times with the Indians. In these skirmishes, the 12th suffered three killed and a like number wounded. The 12th Missouri remained on frontier duty until April 1866 when they were mustered out of service. By May 1866, two thousand three hundred (2,300) rimfire Gallagers were back in storage at the St. Louis Arsenal.[29]

2nd Model Maynard

The wartime production model Maynard carbine arrived in the field in the fall of 1864. By February 1865, the number of Maynards in service was 2,451, of which over 1,600 were in the Military Division of Mississippi.

Military Division of Mississippi[30]
Maynard Carbines
February 1, 1865

6th Indiana.....653	10th Tennessee...184
11th Indiana.....412	12th Tennessee...363

During the early months of 1865, three of the above cavalry regiments sent reports to army headquarters on the efficiency of the Maynard Carbines. Captain Rice of the 10th Tennessee did not have a high opinion of the Maynard when he stated: "For field service it is more ornamental than useful." Lieutenant Charles Sangdom of the 6th Indiana felt that the Maynard was nearly a perfect arm — it was easy to handle and shot very accurately. He and Captain Pratt of the 11th Indiana found one major defect in the Maynard: the tang, which fastened to the stock, was too thin and had a tendency to break with a slight shock in mounting and dismounting. They also felt that the stock was much too light for the gun.[31]

By orders of Irwin McDowell, commanding officer of the Department of the Pacific, the 2nd California Cavalry was directed to turn in their Sharps carbines and be issued Maynard carbines. Companies A, B, G, H and K of the 2nd California located at Camp Union, Sacramento, California, received their Maynards between March 22 and 24. In April, the regiment was supplied with 46,000 Maynard metallic cartridges.[32] During this period, the balance of the 2nd California located in New Mexico and Utah were also issued Maynard carbines.

The only companies of the 2nd California Cavalry to see combat with the Maynards were Compa-

Officers of the 1st New Jersey Cavalry in March 1865. During the Appomattox Campaign, the 1st was armed with Spencer carbines.
U.S. Military History Institute, MOLLUS

nies L and M in a skirmish at Dead Man's Fork, Dakota Territory. At about 10:00 a.m. on June 17, 1865, while resting their horses on a march to overtake a group of hostiles, the 131 troopers led by Colonel Moonlight, of which 87 were from Companies L and M, 2nd California, were attacked by 200 Indians. In the attack, two men from Company L were wounded. The Indians drove off 74 of the cavalry horses, causing most of the men to end up marching the 120 miles back to Fort Laramie on foot.[33] Later in the summer, these same two companies were part of the Powder River Expedition.

Anticipating problems on the Mexican border in which additional cavalry would be needed, Lieutenant Colonel Wainwright, Chief of Ordnance at Benicia Arsenal, requested that an additional 2,000 Maynard carbines plus 500,000 metallic cartridges be sent through Panama to California. A year later, on June 8, 1866, Benicia Arsenal had in storage 4,938 Maynard carbines or twenty-five percent of the total Maynards delivered to the government. Fifty-seven additional Maynard carbines were on hand at the Vancouver Arsenal as of June 1866.

E.G. Lamson (Ball & Palmer)

E.G. Lamson entered into two contracts with the Ordnance Department for 1,000 Palmer carbines and a like number of Ball carbines. The June 1864 contracts called for the carbines to be in .44 caliber with all deliveries by January 17, 1865. Due to the change in caliber from .44 to the new .56-.50 Spencer cartridge, the January 17 deadline could not be met. On May 18, the company requested an additional month to make deliveries on the Palmers. As of this date, 250 Palmers were ready for inspection. The cause for the month extension was that the person entrusted to finish the barrels was found to be incompetent and; therefore, the barrels had to be redone. Of the 2,000 barrels manufactured, over half would have failed ordnance inspections for a variety of reasons. The request was granted and, by the middle of June, all 1,000 Palmer carbines had been received at the New York Arsenal.[34]

With the completion of the Palmers, Lamson requested through Colonel Thornton, on June 6, an extension of eight months to make deliveries on the Ball contract. Lamson needed to make changes to tool-up for the fabrication of the Ball because of the government change in the caliber of the arm. Dyer granted the extension on June 17. Seeing that he was going to come up short on making deliveries, Lamson, on February 22, 1866, requested two more months or until May 17, 1866. The time was granted and, on April 17, the first 500 Ball carbines were sent to the New York Arsenal. By May 24, 1866, the New York Arsenal had in inventory 999 Palmers and 1,000 Ball carbines.[35]

Remington

E. Remington & Sons and their agent Samuel Norris of Springfield, Massachusetts, were granted contracts for 15,000 Remington carbines in .50 caliber and 5,000 in .44 caliber, which was later changed to .46 caliber. With the Remington factory already in full production on government contracts, the carbine contracts were turned over to Samuel Norris. Remington agreed to pay Norris a royalty of three dollars for each carbine delivered to the Ordnance Department. Norris, lacking facilities to manufacture the Remington carbines, contracted with the Savage Revolving Firearms Co. of Middletown, Connecticut for the actual production of the carbines.[36]

In late 1864, the State of New Hampshire was in the process of forming a new state militia cavalry regiment. Needing equipment for the regiment, the state on December 19, 1864, requested from the Ordnance Department 1,200 .44 caliber Remington carbines, 500 Remington revolvers and 1,200 sets of cavalry accoutrements. When the arms were not forthcoming, the state dropped the plans to form the cavalry regiment. Finally, on December 4, 1865, Lieutenant Cullen Bryant, acting Military Storekeeper at New York Arsenal, sent invoices for 1,200 Remington carbines to the New Hampshire Adjutant General's office at Concord. The State of New Hampshire, no longer in need of such a large quantity of carbines, requested that the number sent be reduced to 200 Remington carbines. It appears the State's request was accepted. The Remington carbines were used by the New Hampshire militia during the post-war period.[37] In October 1866, the State of Vermont was listing 1,196 Remington carbines and 482 Smith carbines in storage.

Rimfire Starr

In February 1865, the Ordnance Department placed an order for 3,000 rimfire .56-.50 Starr carbines. The order was increased by an additional 2,000 carbines in April. From the Ordnance Depot at Winchester, Virginia, on March 21, Lieutenant McKee requested that all of the altered (rimfire) Starr carbines be sent there for issue. A week later, 1,000 Starr rimfire carbines were sent from the factory to Winchester on Order for Supplies No. 2,630. A large portion of these rimfire Starrs were issued to the 12th Pennsylvania Cavalry that was on duty in the Winchester area during the spring campaign. The 12th's Starrs were issued prior to Lee's surrender to Grant on April 9. In June, the 12th was listing 671 rimfire Starr carbines in inventory. During this period, the 14th Pennsylvania was mustered out of federal service and turned in 163 Starr carbines, which may have been the rimfire Starr.

On July 12, Everett Clapp, president of the Starr Arms Co., wrote the Secretary of War offering to sell the government an additional 5,000 rimfire Starr carbines at the discounted price of $15 each. These arms had previously been inspected by Ordnance inspectors at the Starr factory. Secretary of War Stanton turned Clapp's letter over to the Chief of Ordnance for his review. From the Ordnance Department came the response that while the rimfire Starr carbines had been issued and the field reports were favorable, no orders should be given since the department currently had on hand 60,000 carbines and another 25,000 carbines under contract; if it were not for this, the department would direct that the rimfire Starrs be obtained by the government. No order was given for these arms.[38] Some of these excess Starrs may have been part of the 250 Starr carbines sold to Canada in 1866.

In May 1865, the St. Louis Arsenal requested that, in addition to the rimfire Gallagers in transit to the Arsenal, the rimfire Starr carbines should also be sent for frontier service. In late July, Major McNutt from the Leavenworth Arsenal requested

The End of Hostilities, 1865

The Windsor Manufacturing Co. stationery head for the Ball carbines c.1865-1866. *National Archives collection*

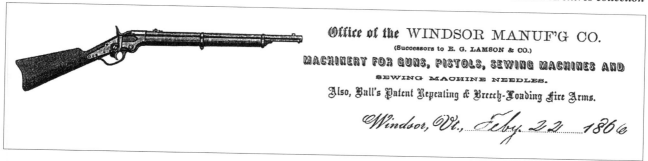

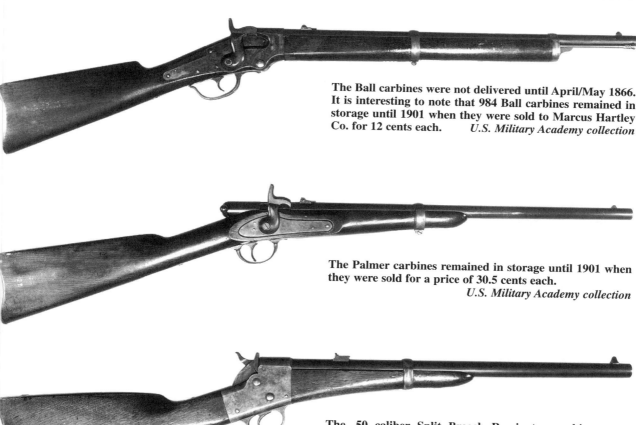

The Ball carbines were not delivered until April/May 1866. It is interesting to note that **984 Ball carbines remained in storage until 1901** when they were sold to Marcus Hartley Co. for 12 cents each. *U.S. Military Academy collection*

The Palmer carbines remained in storage until 1901 when they were sold for a price of 30.5 cents each. *U.S. Military Academy collection*

The .50 caliber Split Breech Remington carbines were received too late for combat use, but a few thousand were sent to the states in the post-war period for state militia use. *U.S. Military Academy collection*

that 300,000 Starr New Model metallic cartridges (.56-.50) be sent from the St. Louis Arsenal. Since St. Louis did not have the stated ammunition, it had to be sent from the New York Arsenal. By the fall of the year, the 2nd U.S. Cavalry was armed with the .56-.50 rimfire Starr carbines. As late as September 1870, 163 rimfire Starr carbines were in storage at the Cheyenne Ordnance Depot.

Triplett & Scott

On January 2, 1865, the State of Kentucky entered into a contract with Meriden Mfg. Co. of Meriden, Connecticut, for 5,000 seven-shot repeating Triplett & Scotts. The price set on these arms was $30 for the 30-inch barrel rifles and $22 for the 22-inch carbines. The contract called for 3,000 of the Triplett & Scotts to have 30-inch barrels and the balance of the order to have the shorter 22-inch barrel carbine length. Kentucky Governor Bramlette intended to use the Triplett & Scotts for the state militia units that were to be called out to help guard Sherman's supply lines against Confederate guerilla and cavalry attacks. The Triplett & Scotts were delivered too late in the year to see field service. Under the Indemnification Act of June 1861, the State of Kentucky filed claims with the federal government seeking reimbursement for the amount spent on these arms. In the fall of 1870, their claim was paid and the Triplett & Scotts were shipped from the state arsenal to the New York Arsenal for storage. Most of them remained in storage until their sale in 1901.[39]

Warner

The first cavalry regiment to be issued the Warner carbine was the 1st Wisconsin. As of September 1864, the regiment was showing 200 Warners, 142 Joslyns, 12 Merrills and 1 Spencer carbine in regimental inventory. By February 1865, the number of Warners in the 1st Wisconsin had increased to 341 carbines. Early in March, they exchanged their Warners for the Spencers that they used on Wilson's Selma raid. Before the regiment turned in their Warners, they reported that the Warner worked well and was very accurate, but that the springs and the small of the stock were too weak and easily broken.

In January 1865, the 3rd Massachusetts Cavalry, while at Remount Camp, Pleasant Valley, Maryland, was issued the Warner carbine for field trials. The field reports of January 31 to Major Theodore C. Otis, Chief of Ordnance, Army of the Shenandoah, stated that the accuracy of the Warner carbine was astounding and comparable to the Sharps issued to the 3rd. They did, however, find major difficulty in loading and extracting the rimfire cartridge from the chamber. Major David T. Bunker, Company C, said he had found the following problems with the 26 Warners assigned to his company:

1. The cartridge had to be pushed in with a screwdriver.
2. It required a severe blow of the butt on a log to throw the plunger back to open the chamber.
3. The breechblock required a severe blow with a stick to open it after firing.
4. A screwdriver was required to remove the cartridge case.
5. Other problems were the breechblock blowing off; the stock breaking off; and the breech pins breaking.

The complaints from Company F indicated that not one Warner carbine could be fired 20 times without it failing. Because of the negative reports, the Warner was sent to the Washington Arsenal for further examination. The 3rd Massachusetts was then issued over 500 Sharps carbines for their spring campaign in the Shenandoah Valley.

From the Washington Arsenal on February 20th, J.S. Dudley reported to Major Benton, commanding officer at the arsenal, that the problem with the 3rd Massachusetts' Warners was the ammunition. The cartridges manufactured by Crittenden and Tibbals lacked fulminate to ignite them and the copper cases were of a very inferior quality. A problem not stated in Dudley's report was that the Warner cartridges were smaller than the Spencer cartridges issued to the regiment. This was the real problem behind the failure of the Warners. In the post-war period, the Ordnance Department would have 2,000 of the Warner carbines rechambered to use the .56-.50 Spencer cartridge.[40]

Major Callendar, from the St. Louis Arsenal, notified the Ordnance Department on February 21 that he had received Orders for Supplies No. 1,494 for the issue of 500 sets of accoutrements for the Warner carbine. The request had been received from Governor John Evans of Colorado Territory at Denver. It appears that the Warner was issued to the 1st Colorado Cavalry for use against the Indians. A year later, in May 1866, the St. Louis Arsenal was

The End of Hostilities, 1865

Starr Arms Company stationery head c.1865. In July, the company had offered the government 5,000 Starr carbines at a reduced price of $15.00 each. This offer was turned down due to the large quantities of carbines on hand or on order.
National Archives collection

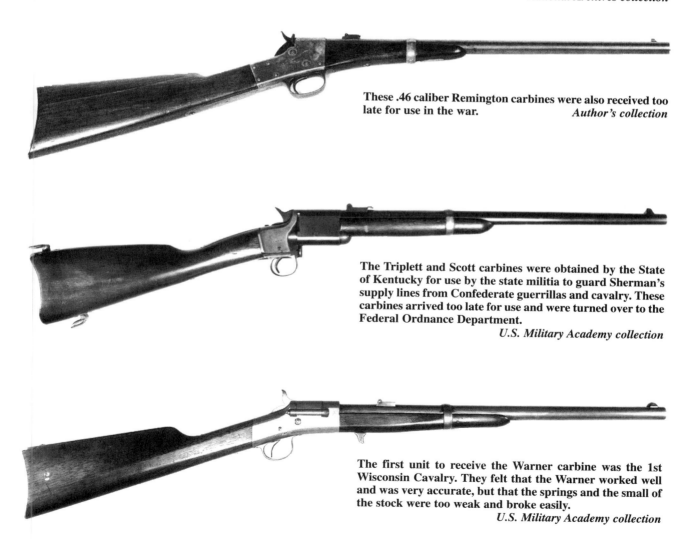

These .46 caliber Remington carbines were also received too late for use in the war. *Author's collection*

The Triplett and Scott carbines were obtained by the State of Kentucky for use by the state militia to guard Sherman's supply lines from Confederate guerrillas and cavalry. These carbines arrived too late for use and were turned over to the Federal Ordnance Department.
U.S. Military Academy collection

The first unit to receive the Warner carbine was the 1st Wisconsin Cavalry. They felt that the Warner worked well and was very accurate, but that the springs and the small of the stock were too weak and broke easily.
U.S. Military Academy collection

listing 321 Warner carbines in their inventory.

Spencers for the 6th U.S. Cavalry

During the late part of the war, only the 4th U.S. Regular Cavalry, located in the western Theater of operations, was armed with Spencer carbines. The Spencer carbines had been issued to the 4th U.S. in early 1864. The balance of the U.S. regulars were armed with Sharps carbines during the 1864-1865 time period.

The second regiment of regular cavalry to receive Spencer carbines and the first unit to obtain the Model 1865 Spencer chambered for the .56-.50 cartridge was the 6th U.S. Regular Cavalry. In the summer of 1865, Brevet Lieutenant Colonel Morris, commanding officer of the 6th U.S., requested 853 Spencer carbines for the regiment. During August and September, while located near Frederick, Maryland, the 6th U.S. was issued 30 Model 1860 Spencer carbines, 803 Model 1865 Spencers and 20 Sharps carbines. The ammunition issued to the regiment contained the old cartridge for the wartime model Spencer, not the Model 1865. Therefore, on October 6, Major S.H. Starr of the 6th requested that the correct ammunition for the Model 1865 be issued; this was quickly done by the Ordnance Department.[41]

Footnotes — The End of Hostilities, 1865

[1] NARG 156-21
[2] Ibid.
[3] Stephen Z. Starr, *The Union Cavalry in the Civil War, Vol. II*, Baton Rouge, LA, 1981, pp. 428-431
[4] Ibid., p. 428 and NARG 108-77
[5] ORs Series I, Vol. 46, Part I, pp. 1135-1136
[6] Ibid., p. 1118
[7] Ibid., p. 1136
[8] Stephen Z. Starr, op.cit., p. 463
[9] ORs Series I, Vol. 46, Part I, pp. 1132 and 1136
[10] Stephen Z. Starr, op.cit., p. 472
[11] NARG 156-21
[12] NARG 156-21, February 1, 1865 Monthly Report of Field Artillery and Small Arms in the Military Division of Mississippi & OR Series I, Vol. 49, Part I, pp. 402 and 403; 799 and 800
[13] OR Series I, Vol. 49, Part I, p. 597
[14] NARG 156-21, Monthly Reports for February and March for the Department
[15] OR Series I, Vol. 49, Part I, p. 737
[16] ORs, Ibid., p. 808 and Part II, p. 19
[17] Ibid., Part I, pp. 438 and 448
[18] Stephen Z. Starr, Vol. I, op.cit., pp. 42 and 43
[19] NARG 108-77
[20] NARG 156-21, Box 243
[21] Ibid., Box 246 and NARG 108-77
[22] OR Series I, Vol. 44, Part I, p. 361
[23] NARG 156-21, Box 241, Letter from Kilpatrick dated March 28, 1865
[24] OR Series I, Vol. 47, Part I, p. 890
[25] NARG 156-21
[26] NARG 156-21, Department of Virginia Monthly Report- March 1865 and OR Series I, Vol. 46, Part II, p. 582 and Part III, p. 1108, plus Frederick H. Dyer, *A Compendium of the War of Rebellion*, Dayton, OH, 1978, p. 1386
[27] Annual Report Quartermaster General to the Governor of the State of Kentucky for the year 1867, Frankfort, KY, 1867, p. 12
[28] NARG 156-21, Box 244 and OR Series I, Vol. 48, Part II, p. 633
[29] NARG 156-21, Box 252, Frederick H. Dyer, op.cit., pp. 1023, 1109, 1311 and 1641
[30] NARG 156-21, Miltary Department of Mississippi February Report
[31] NARG 108-75
[32] NARG 156-21
[33] OR Series I, Vol. 48, Part I, pp. 325-328
[34] NARG 156-21 Box 241
[35] Ibid., Boxes 241, 249 and 253
[36] Roy M. Marcot, "Remington Split-Breech Carbines," *The Gun Report*, February 1991, pp. 14-20
[37] NARG 156-21, Boxes 243 and 246
[38] Ibid., Box 247
[39] Andrew F. Lustyik, "Triplett & Scott Carbine, Part II," *The Gun Report*, July 1979, pp. 12-20
[40] Ibid., "The Warner Carbine," *The Gun Report*, May 1960, pp. 6-9
[41] NARG 156-21, Box 244

GOVERNMENT ◆ DISPOSALS ◆

Within five months of Lee's surrender in April of 1865, the Ordnance Department started to dispose of its vast quantities of carbines in arsenal storage and those being turned in by the cavalry as they were mustered out of federal service. Over the next 40 years, more than 200,000 carbines were sold on the market by the government.

The first sales of these carbines occurred at Harpers Ferry, West Virginia, in September of 1865 where the following unserviceable arms were sold:

Sales at Harpers Ferry[1]
September 1865

Qty.	Item	Amount Received
233	Smith Carbines	$183.35
255	Starr Carbines	191.25
39	Gallager Carbines	8.10
2	Joslyn Carbines	.60
2	Gibbs Carbines	.40

A second public sale was held at the same location on Tuesday, February 27, 1866, with only 24 Gallagers, 1 Remington and 4 Burnside carbines sold. The items withdrawn from the sale due to the lack of a minimum bid were:

Ltd. Bid	Qty.	Item	Condition[2]
$10.00	1,270	Joslyn cal. 52	New
10.00	352	Starr cal. 54	New
4.00	1,909	Burnside cal. 54	like sample
3.00	472	Burnside cal. 54	like sample
Any Price	214	Burnside cal. 54	like sample & 200 barrels

A month before, on January 18, the Allegheny Arsenal sold 1,417 carbines to the public for a total of $878.80. Carbines sold consisted of 243 Gwyn and Campbells, 59 Merrills, 505 Smiths and 610 Gallagers. Items not sold due to lack of bids were:

Qty.	Item	Minimum Price[3]
67	Ballard Carbines	$3.00
632	Burnside Carbines	3.00
538	Joslyn Carbines	3.00
444	Sharps Carbines	4.50
64	Starr Carbines	3.00
27	Wesson Carbines	4.00
220	Rifled Pistol Carbines	1.50 to 2.00

At the direction of the Ordnance Department, the 67 Ballards were sent to the New York Agency in February.

In 1867, nearly 90 percent of all carbines sold were from the November sale at the Leavenworth Arsenal.

Sales at Leavenworth Arsenal[4]
November 1867

Carbines	Price Ea.	Purchaser
1,349 Gallager	$2.00	Schuyler, Hartley & Graham
348 Hall	.50	"
199 Joslyn	5.00	"
149 Joslyn	.50	"
1,958 Starr	4.00	"
408 Starr	1.00	"
22 Wessen	2.00	O.H. Viergertz & Co.
57 Maynard	1.25	Wm. Syms & Bros.
34 Smith	1.00	"

During the later part of the 1860s, nearly half of all carbines sold were Starr carbines. By the end of 1869, nearly 13,000 Starrs had been sold.

Sales of Starr Carbines[5]
1865 - 1869

Year	Quantity		Price	Purchaser
1865	266	Percussion Starr		
1866	254	Percussion Starr		
Oct. 1867	200	Rimfire Starr	$10.00	I.F. DeSilver & Son
Nov. 1867	1,958	Percussion Starr	4.00	Schuyler, Hartley & Graham
Nov. 1867	408	Percussion Starr	1.00	Schuyler, Hartley & Graham
Dec. 1867	40	Percussion Starr	10.00	I.F. DeSilver & Son
Dec. 1867	20	Percussion Starr	6.25	Wm. Syms & Bros.
June 1868	187	Starr	6.00	W.F. Weld
June 1868	272	Starr	3.60	C.W. Pond
June 1868	105	Starr	2.25	C.W. Pond
July 1868	610	Starr	5.50	Wm. Read & Sons
Aug. 1868	2,180	Rimfire Starr	10.00	W.W. Walcott
Aug. 1868	100	Rimfire Starr	10.00	H. Boker & Co.
Aug. 1868	728	Percussion Starr	10.00	W.W. Walcott
Sept. 1868	1,000	O.M. Starr	5.60	Schuyler, Hartley & Graham
Oct. 1868	180	O.M. Starr	10.00	H. Boker & Co.
Nov. 1868	2,698	O.M. Starr	10.00	W.W. Walcott
Nov. 1868	554	N.M. Starr	10.00	W.W. Walcott
Nov. 1868	200	O.M. Starr	5.60	C.H. Townsend
Dec. 1868	299	Starr	$1.60 – $4.10	Charles Folson
April 1869	374	Starr	1.00 – 2.00	
April 1869	150*	N.M. Starr	9.60	Wm. Read & Son
April 1869	20	O.M. Starr	5.60	W.F. Weld
Total	12,803			

* These 150 NM Starrs had been received from the Allegheny Arsenal in April.

Up through 1868, the Ordnance Department had sold less than 100 Spencer carbines. On December 26, 1868, the Consulate of Argentina in New York City wrote to the Secretary of War requesting to purchase 200 Spencer carbines and 1,000 cartridges per arm. Approval was given on January 2, 1869, and the arms were delivered from the New York Agency.

Sales of Spencer Carbines to Argentina[6]

Date	Item Sold	Price	Total
1/2/69	200 M1865 Spencer Carbines with Stabler cut-off	$26.00	$5,200.00
	200,000 cartridge .56—50	$18.00/1,000	3,600.00
		Total	**$8,800.00**

Government Disposals

The San Antonio Arsenal, in June 1869, requested approval to sell the 753 caliber .52 Spencer carbines on hand at the arsenal. At that time, on the San Antonio open market, second-hand Spencers were selling for from $18 to $20 apiece with the ammunition at $35 per 1,000. Many of the Spencer carbines locally were being sold to the Mexican market.[7] There is no record of any sales being made from the San Antonio Arsenal in 1869.

The St. Louis Arsenal, in January 1869, sent to Colonel Crispin at the New York Agency over 32,000 serviceable carbines for sale — all fourth class, of which there were about 12,000 remaining at St. Louis. These 4th class carbines either had broken stocks, missing parts or barrels that had burst. On April 12 and 13 of that year, these carbines were placed on sale at auction. Over 3,100 carbines were sold as shown in the following schedule:

Ordnance Sales at St. Louis Arsenal[8]
April 12-13, 1869

Lot No.	Items	Quantity	Sales Price per Item
208	Starr	234	$2.00
209	Starr	140	1.00
210	Maynard	118	2.30
211	Maynard	198	1.00
213	Sharps	562	2.25
214	Wesson	5	1.00
215	Wesson	20	.35
216	Symmes	9	2.00
218	Warner	50	.50
222	Gallager	1,139	.50
226	Joslyn	99	1.50
230	Ballard	297	3.00
231	Ballard	156	.75
234	Gibbs	58	.50
235	Sharps & Hankins	8	.30

In addition to the above arms, lot no. 315 included 46 attachable stocks for the pistol carbines. These stocks were sold for twenty-five cents each.

1870 Sales

The outbreak of the Franco-Prussian War in July 1870 led to French demands for quality small arms, providing U.S. arms dealers with a ready market for surplus weaponry. It also gave the Ordnance Department a local market for sales of their surplus small arms.

One of the first dealers to approach the Ordnance Department was E. Remington & Son for the purchase of Spencer cartridges. On September 20, the New York Agency sold Remington 400,000 Spencer cartridges at a cost of $18 per thousand.[9] A few days later, Schuyler, Hartley & Graham offered to pay $18 each for 5,000 of the M1865 Spencer carbines. On September 30, they increased their offer for all the Spencer carbines then on hand at the arsenal. During October, Schuyler, Hartley & Graham obtained 5,000 caliber .50 M1865 Spencer carbines from New York at $20 each.

At the beginning of October, the New York Agency intended to sell a maximum of 15,000 Spencer carbines. This total was quickly surpassed by the overwhelming demand. By the end of the month, nearly 70 percent of the entire current inventory of Spencer carbines had been disposed of at prices ranging from $20 to $25.25 per arm. In total, 34,438 Spencer carbines were sold to the following firms and individuals:

Sale of Spencer Carbines[10]
New York Agency
October 1870

Quantity	Description	Price	Purchaser
5,000	M1865 Spencer cal. 50	$20.00	W.S. Starr
4,600	M1865 Spencer cal. 50	23.00	O.F. Winchester
400	M1865 with Stabler	23.00	O.F. Winchester
5,000	M1865 Spencer cal. 50	20.00	Schuyler, Hartley & Graham
7,019	O.M. Spencer cal. 52	25.25	E. Remington & Son
12,419	M1865 (12,160 w/Stabler)	25.25	E. Remington & Son

The Spencer ammunition was also in great demand. Between September and the end of the year, the New York Agency sold a little less than 30 million Spencer cartridges at prices ranging from $16 to $18.25 per thousand. Most of the ammunition was set at $16 per thousand. The sales breakdown follows:

Disposal of Spencer Ammunition[11] from New York Agency September - December 1870

cal. .52	cal. .50	Purchaser
11,955,678	6,166,944	E. Remington & Son
1,201,521	2,811,999	Schuyler, Hartley & Graham
806,400	2,016,000	W.S. Starr
	2,520,000	O.F. Winchester
1,332,576		Other Dealers
15,296,175	13,514,943	Total

On October 17, 1870, General Dyer, Chief of Ordnance, offered a large quantity of Ordnance stores for public sale by sealed bids. Bids would be received until 2:00 p.m. on Saturday, October 29. All successful bids would require 20 percent down payment at the time of the award and the balance upon delivery, which was promised within 30 days thereafter. In addition to the revolvers, cavalry sabres and breechloading .58 caliber muskets, the following carbines and appendages were up for bids:[12]

- 1,000 Ball's repeating carbines, cal. 50, with ammunition
- 2,500 Gallager's carbines, adapted for Spencer ammunition
- 4,000 Maynard's carbines with 500 rounds of ammunition per arm
- 1,000 Palmer's carbines, caliber .44, with ammunition
- 3,600 Remington carbines, caliber .44 with ammunition
- 2,500 Warner's carbines and ammunition
- 2,700 Joslyn's carbines and ammunition, caliber .52
- 40,000 sets carbine accouterments, serviceable, Blakeslees and other patterns

On Friday, October 28, the following sealed bid was received from Schuyler, Hartley and Graham:[13]

New York October 28, 1870
Gen. A.B. Dyer
Chief of Ordnance
Washington
D Sir-

Agreeable to your advertisement, we make the following offer — for 2500 Gallager Carbines new adapted for Spencer Ammunition 50 or 53 Cal — $12.25

2500 Warners carbines new ditto — 12.25

1000 Spencer Rifles with bayonets new $25.50 with privilege of 400 Rounds of Ammunition for each arm if we desire it at $18 per thousand.

If this offer is not promptly accepted we do not wish you to consider it binding upon us, as we shall in that case withdraw it.

It is understood, the goods are to be delivered at wharf in New York — no charge for boxes or accouterments.

Yours Truly

(signed) Schuyler Hartley Graham

The Ordnance Department accepted their offer for the 2,500 Gallagers and Warner carbines but they were out-bid on the Spencer rifles.

Four hundred cartridges were to be supplied with each of the Warner carbines. Schuyler, Hartley and Graham were unaware that the Warner took special ammunition. When the Ordnance Department went to supply the Warner ammunition, the New York firm refused it since it was not acceptable for any other carbine. The Warners sold to the firm were chambered for the Warner cartridge with a large portion chambered for the .56-50 Spencer cartridge. Finally, they agreed to accept the Warner ammunition at $16 per thousand with 925,372 rounds being received. The Warner ammunition was sent from the St. Louis Arsenal.[14]

No sales were made at this time for the 1,000 Ball, 1,000 Palmer and 4,000 Maynard carbines. These arms remained in the federal arsenals for another 30 years. The Model 1864 Joslyn carbines were sold in two lots. The first lot of 160 Joslyn carbines was purchased by W.S. Starr in October and the second lot of 2,600 Joslyns was obtained by Austin Baldwin & Co. in December. Both purchases were obtained at the sale's price of $12.50 each. In November, E. Remington and Son purchased the largest quantity of carbines when they took receipt of 14,757 Remingtons; 7,487 of these carbines had been shipped the previous month from the Springfield Arsenal to New York. Most of the Remington carbines sold were of .50 caliber with a few .44 calibers also sold. Even after the Remington purchase, the New York Arsenal still had 310 Remington .50 caliber carbines on hand. These arms were a little rusty but could be made like new with very little work.[15]

1871-1900

When the State of Kentucky turned over their Ballards and Triplett & Scotts to the federal government, they were received at the Indianapolis Arsenal. In October 1871, these arms were transferred to the New York Agency where they arrived on October 21.

State of Kentucky[16]
Small Arms Received at New York Agency
October 21, 1871

	New	Clean & Repaired	Unserviceable	Total
Ballard Rifles cal. 46	685	688	485	1,858
Ballard Carbines cal. 44	100	996	645	1,741
Triplett & Scott – 22" barrel	1,900		93	1,993
Triplett & Scott – 30" barrel	2,810	50	133	2,993

By the 1870s, the percussion carbines in the various arsenals were totally obsolete and the prices received reflected their value on the arms musket. The following are examples of these sales:

Percussion Carbine Sales
1871-1876

April 1873	1,009 Burnside	$0.105 each
October 1871	177 Smith	.15 each
October 1871	1,165 Merrill	.15-.65 each
February 1872	66 Joslyn	.50 each
June 1874	309 Smith	.17 each
June 1875	50 Starr	.25 each
October 1876	209 Starr	.85 each
October 1876	68 Joslyn	.43-.575 each
October 1876	918 Smith	.25 each
October 1876	1,571 Merrill	.22 each
June 1875	6 .54 cal. Perry	.22 each

The major purchaser in the 1875-76 time period was the firm of H. Boker & Co. During this period, they purchased over 4,000 Spencers and a like quantity of Remington carbines as shown in the following schedule:

April 1875–January 1876	3,663 Remington Carbines .46 caliber
February 1876–March 1876	382 Remington Carbines .50 caliber
July 1876–November 1876	2,058 1st Class Spencer Carbines
July 1876–November 1876	1,577 2nd Class Spencer Carbines
July 1876–November 1876	400 3rd Class Spencer Carbines

H. Boker & Co. also purchased 32 Triplett & Scotts at $6.00 each during this period.

The 1880s and 1890s were a slow period for sales of the old Civil War carbines. The one exception occurred in October 1882 when all of the Kentucky Ballard carbines were sold for prices ranging from $1.50 to $4.00 each.

Sale of Ballard Carbines[17]
October 1, 1882

385 Ballard Carbines	cal.42	$3.00 ea.	Schuyler Hartley & Graham
149 Ballard Carbines	cal.42	2.75 ea.	Schuyler Hartley & Graham
559 Ballard Carbines	cal.42	1.50 ea.	Schuyler Hartley & Graham
104 Ballard Carbines	cal.42	2.00 ea.	Schuyler Hartley & Graham
142 Ballard Carbines	cal.44	4.00 ea.	Schuyler Hartley & Graham
265 Ballard Carbines	cal.44	2.00 ea.	Schuyler Hartley & Graham

Total 1,604

1901 Sales

At the turn of the century, nearly half of all the carbines in storage as of 1866 were still in storage. In 1901, the government made a constituted effort to clear their inventory of these obsolete arms. By year's end, over 90,000 carbines had been disposed of. Two thirds of the entire year's sales occurred in June when over 62,000 carbines were sold from the New York Agency.

Sales From New York Agency[18]
June 1901

Kirkland Bros.

2,349	Triplett & Scott	$.25 each
545	Triplett & Scott	.10
1,811	Triplett & Scott	.20

Marcus Hartley Co.

72	Triplett & Scott	$.21-.30 each
984	Ball	.12
495	Ballard .44 cal.	.86
475	Ballard .54 cal.	.12-.21
17,109	Burnside	.06
1,376	Hall .52/54 cal.	.06
2,096	Joslyn M1862/64	.06-.275
24	Joslyn M1855	.12-.35
1,935	Merrill	.11-.26
7,995	Smith	.04-.08
743	Spencer .52 cal.	.35-.65
4,364	Starr .54 cal.	.06-.26
3,127	Sharps .52 cal.	.06-.32
745	Sharps .50 cal.	.21-$1.05

Francis Bannerman

109	Sharps .52 cal.	$0.06-.27 each
5,320	Burnside	0.0317-.0755
5,597	Gallager	0.0527-.1627
192	Gibbs	0.0455-.0727
130	Hall .64 cal.	0.1127
3,379	Maynard	0.1327-.1758
1,414	Merrill	0.0527
40	Starr	0.0309
225*	Lindsay	
	Rifle Musket	0.3155
9,999*	Remington M1863 Rifle w/Sword Bayonet	0.5357

While not carbines, these arms are of general interest to the Civil War collector.

The carbine ammunition sold at this sale was:

446,169	Sharps	52 cal. linen cartridges	Price not stated	
5,880	Starr	.54 cal. linen cartridges	$1.537	per thousand
57,800	Ballard	.42/44 cal. cartridges	.76-.80	"
1,597,942	Spencer	56-50 cal. cartridges	1.21	"
676,936	Spencer	56-52 cal. cartridges	1.14-1.27	"
420,000	Burnside	cal. .54 cartridges	1.51	"
200,000	Gallager	cal. .51 cartridges	.83	"
412,000	Maynard	cal. .50 cartridges	1.11	"
37,800	Wesson	cal. .44 cartridges	.57	"

Later in the year, large quantities were sold at various other arsenals as shown in the following chart:[19]

Kenaba Arsenal Sales September 1901

	Items Sold	Price Per Item	Purchaser
7,954	Smith Carbines w/1,194 lbs. of Spare Parts	$0.162	Nolan Brothers
3,321	Maynard Carbines w/99 lbs. of Spare Parts	0.18	M. Hartley Co.
999	Palmer Carbines w/50 lbs. of Spare Parts	0.305	M. Hartley Co.

Benicia Arsenal Sale November - December 1901

5,678	Maynard Carbines - New	$0.1925	W. Stokes Kirk
1,512	Spencer Carbines .52 cal.	0.68-$1.57	Kirkland Bros.
523	Spencer Carbines .50 cal.	0.266-.486	Francis Bannerman
972	Sharps Carbines 50/70	1.00	Sears Roebuck Co.
547	Sharps Carbines 50/70	0.128	M. Hartley Co.
2,489	Sharps Carbines .52 cal.	.076-.328	Francis Bannerman

Allegheny Arsenal November 1901

768	Burnside Carbines	$.31-.4188 ea.	Nathan Spering
53	Starr Carbines	.35	C.J. Godfrey
1,213	Gwyn & Campbell Carbines	.3888	Francis Bannerman
83	Gallager Carbines	.3888	Francis Bannerman
504	Joslyn Carbines	.3888	Francis Bannerman
325	Lindner Carbines	.4188	Francis Bannerman
603	Smith Carbines	.4188	Francis Bannerman

By the end of 1901, the arsenals had nearly sold out their inventory of carbines. The next few years would see continued sales of a few small quantities of arms. The story of the Civil War carbines for all practical purposes, had come to a close with the ordnance sales of 1901.

Footnotes — Government Disposals

[1] NARG 156-125
[2] NARG 156-21 Box 251
[3] Ibid.
[4] NARG 156-125
[5] NARG 156-124
[6] Ibid., and 156-21 Box 275
[7] NARG 156-21 Box 274
[8] Ibid.
[9] Ibid., Box 280
[10] NARG 156-125
[11] NARG 156-124
[12] NARG 156-21 Box 278
[13] Ibid., Box 280
[14] Ibid., Box 282 and NARG 156-124
[15] Ibid., Box 280 and NARG 156-125
[16] NARG 156-6
[17] NARG 156-124
[18] Ibid.
[19] Ibid.

During the Modac War, the Indian scouts are shown with Spencer carbines while the Army was armed with both the Sharps and Spencer.
National Archives collection

The Indian Wars 1866-1891

A 10th Cavalry, Company D Sergeant armed with a .50-70 Sharps carbine. This photo was taken at Fort Sill c.1871. In September 1871, the 10th had 845 Sharps carbines in inventory.
The Larry Jones collection

POST-WAR PERIOD
◆ 1866-1873 ◆

The year 1866 started with nearly 30 volunteer cavalry regiments and independent cavalry battalions still in federal service, mostly in the West guarding against Indian outbreaks. By July, the number of volunteer regiments had decreased to a half dozen. Before the volunteers were mustered out in 1866, a few of the units were armed as follows:[1]

> 2nd California Cavalry — 120 Joslyn, 416 Sharps
> 1st New Mexico Cavalry — 142 Sharps
> 11th Ohio Cavalry — 118 Spencer, 33 Joslyn
> 1st Nebraska Cavalry — 471 Joslyn
> 3rd U.S. Colored Cavalry — 850 Joslyn
> 5th U.S. Colored Cavalry — Joslyn
> 6th U.S. Colored Cavalry — 1,000 Gwyn & Campbell

The regular cavalry was armed as of June 1866 with:

U.S. Regular Cavalry[2]
June 30, 1866

	Sharps	Spencer	Maynard	Starr
1st U.S.	778	2	121*	
2nd U.S.		133		436
3rd U.S.	800			
4th U.S.		674		
5th U.S.	867			
6th U.S.	48	347		
Total	2,493	1,156	121	436

* Issued to Companies B and L

For the fiscal year ending on June 30, 1866, the Leavenworth Arsenal had issued to the various cavalry units over 1,200 carbines. The arms included 66 Burnsides, 500 Model 1864 Joslyns, 48 Sharps, 209 Spencers, 342 rimfire Starrs and 50 Merrill carbines. Fort Union Arsenal in Arizona had over the past fiscal year issued the following ordnance to the troops:[3]

Sharps Carbines.........................111
Sharps Cartridges....................35,000
Rifle Muskets............................62
Rifle cal. .58...........................50
Rifle cal. .54...........................15
Colt Pistols cal. .44....................31
Colt Pistols cal. .36....................12
Remington Pistols .44...................187

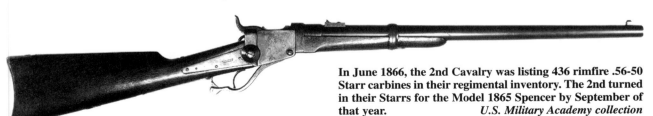

In June 1866, the 2nd Cavalry was listing 436 rimfire .56-50 Starr carbines in their regimental inventory. The 2nd turned in their Starrs for the Model 1865 Spencer by September of that year. *U.S. Military Academy collection*

The following four charts reflect the quantities of carbines that were in storage at the various arsenals and ordnance depots as of June 1866. Nearly 210,000 carbines were listed as being on hand at these locations.

Carbines In Storage[4]
June 1866

Location	Ball	Palmer	Ballard	Burnside	Cosmopolitan
Allegheny				918	1,216
Baton Rouge				4,944	
Benicia				1	
Charleston				182	
Columbus				360	47
Frankford			400	132	
Hilton Head				511	
Kenneche				31	
Leavenworth				899	529
Fort Monroe				10	
Mt. Vernon				566	17
New York Arsenal	1,000	999	99	46	
New York Agency			4	1,646	146
Springfield			6	93	
St. Louis			347	8,045	3,040
Washington			448	2,820	
Watertown				249	
Watervliet				820	
Total	**1,000**	**999**	**1,304**	**22,273**	**4,995**

	Gallager					
Location	OM	NM	Gibbs	Joslyn	Lindner	Hall
Allegheny	90				342	
Baton Rouge	2		34			
Benicia						158
Charleston						55
Columbus	50					
Frankford	350	2,500		7		
Leavenworth	1,396			1,355		346
Fort Monroe						1
Mt. Vernon	24		21			26
New York Arsenal				1		1
New York Agency	2			3	2	138
Springfield	3		2	88		2
St. Louis	5,201	2,321	238	1,077		495
Washington	754		223	1,414		
Watertown				1		
Watervliet	26			3		5
Total	**7,898**	**4,821**	**518**	**3,949**	**344**	**1,227**

Carbines In Storage
June 1866

Location	Maynard	Merrill	Musketoon	Remington	Sharps	Sharps & Hankins
Allegheny	50	12			571	
Baton Rouge	408	3		1,000	2,217	54
Benicia	4,938		1,293		16	
Charleston					82	
Columbus					47	
Detroit		115			255	
Frankford	1	32	52		1,134	
Hilton Head					373	
Indianapolis	256				296	
Leavenworth	333	1,791			2,380	
Fort Monroe	410	70			245	309
Mt. Vernon		29			1,434	
New York Arsenal	3,323	1	5	16,594	1	
New York Agency	683	14		7	1,612	72
Springfield		3	14	3	1,341	18
St. Louis	3,217	3,819	287		8,675	190
Vancouver	369		154		31	
Washington	67	286			4,954	157
Watertown	240				339	
Watervliet	105	841			1,875	
Total	**14,400**	**7,016**	**1,805**	**17,604**	**27,878**	**800**

Location	Smith	Spencer OM	Spencer NM	Starr OM	Starr NM	Warner	Wesson
Allegheny	603	3,009	739	216			22
Baton Rouge	11	352	1,264	25			
Charleston		250					
Columbus	10	20,397					
Detroit		253					
Frankford	23	898		444			
Hilton Head		95					
Indianapolis		209					
Kenneche			3,000				
Leavenworth	1,693	259	276	3,077			27
Fort Monroe		279		94			
Mt. Vernon	41	699		48			
New York Arsenal	7,955	1	16,164	2	2,000	2,501*	
New York Agency	133		817	334		2	2
Springfield	74	494	2		1,857		
St. Louis	5,219	2,998	250	2,787		321	
Washington	1,591	1,193		2,677			
Watertown		207		11			
Watervliet	125	900		813			
Total	**17,478**	**32,493**	**22,512**	**10,528**	**3,857**	**2,824**	**51**

*2,000 of the Warner carbines in storage at the New York Arsenal were chambered for the .56-.50 Spencer cartridge.

In September 1866, General Grant ordered the Ordnance Department to close several of the southern ordnance depots since there was no longer a need for them. Starting in October, the Hilton Head; Macon, Georgia; Baton Rouge and Harpers Ferry ordnance depots were ordered closed and their ordnance stores shipped to the northern arsenals for storage.

Other transfers occurred during 1866. In February, the Allegheny Arsenal sent 67 caliber .44 Ballard carbines to the New York Arsenal. In June, Frankford Arsenal had about 900 Old Model Spencer carbines in storage; by December, the arsenal had on hand 10,000 Model 1865 Spencer carbines, of which 650 were equipped with the Stabler cut-off. Also in December, Allegheny Arsenal sent to Springfield 3,017 Spencer carbines and 417 Spencer rifles to be altered to the .56-.50 Spencer cartridge. By June of the following year, Springfield had altered 4,383 Spencer carbines and 442 Spencer rifles to the new Spencer cartridges.[5]

With the need for a larger regular army to maintain order on the frontier, Congress, on July 28, 1866, authorized four additional cavalry regiments — the 7th through 10th. The last two regiments, the 9th and 10th Cavalry, were made up of Negro enlisted men and commanded by white officers. These four regiments were organized and armed as shown in the list below:

Regt.	Location	Arms Issued[6]
7th U.S.	Fort Riley, Kansas	Spencer Carbines, Remington Revolvers
8th U.S.	Angel Island, California	Spencer Carbines, Colt Revolvers
9th U.S.	Greenville, Louisiana	Spencer Carbines, Remington Revolvers
10th U.S.	Fort Leavenworth, Kansas	Spencer Carbines, Remington Revolvers

The Spencer carbines issued to the 8th Cavalry came from the Benicia Arsenal. Benicia had requested 1,000 Spencer carbines from the New York Arsenal in December 1865 and had received them on July 7, 1866. These Spencers were the first Spencers to arrive on the West Coast.[7] The 8th U.S. was initially assigned to the Arizona Territory in 1867.

At the end of August 1866, Brevet Colonel McNutt, commanding officer at Leavenworth Arsenal, was directed by orders from General Sherman to issue the Spencer carbine to the two new regiments being raised there (Custer's 7th and Grierson's 10th) and to rearm the 2nd Cavalry with Spencers. McNutt was given the option to issue the Remington carbine to one company in each regiment if he felt that they should be placed in field service. The reason given for rearming the 2nd was that the improved Starr carbines they had been issued, though excellent arms, had two or three weak parts that rendered them unserviceable.[8] McNutt issued to these three regiments 2,280 Spencer carbines .50 caliber, 1,550 Remington revolvers and 2,600 light cavalry sabres.

On the 27th of September, the following quantities of Spencer carbines were issued to the 2nd Cavalry.

Issuance of Spencer Carbines[9]
2nd Cavalry
September 27, 1866

Location	Co.	
Fort Kearny	Co. "F"	90 Spencer Carbines M1865/9,072 cartridges
Fort Laramine	Co. "E"	95 Spencer Carbines M1865/9,072 cartridges
Fort Kearny	Co. "D"	90 Spencer Carbines M1865/9,072 cartridges
Fort Laramine	Co. "H"	95 Spencer Carbines M1865/9,072 cartridges
Fort Kearny	Co. "K"	90 Spencer Carbines M1865/9,072 cartridges
Fort Kearny	Co. "L"	40 Spencer Carbines M1865/4,032 cartridges
Total		500 Spencer Carbines M1865 49,392 cartridges

In November 1866, sixty members of Company C, 2nd Cavalry, arrived at Fort Phil Kearny armed with the .56-.50 Starr carbine and muzzleloading rifle muskets. Four days before Christmas, Captain William J. Fetterman's relief column, consisting of 49 infantrymen armed with muzzleloading rifle muskets, 27 cavalrymen of C Company armed with Spencers and two civilians armed with Henry rifles and three officers were ambushed near the fort by over 1,500 Indians. Captain Fetterman's command was wiped out to a man. This action has come to be known as the Fetterman Massacre.[10]

1867-1869

In December 1866, Lieutenant Colonel George Crook arrived at Fort Boise, Idaho, to take overall command of the local military district of operation. Over the next year and a half, he would lead the 1st Cavalry and 14th Infantry against the Paiute Indians. Crook's aggressive approach to fighting the Paiutes, which included using other Indians on his side, forced them to request peace in July of 1868. In over 40 engagements, the Paiutes suffered nearly 600 casualties. During most of this period, the 1st cavalry was armed with percussion Sharps and Maynard carbines.

The 9th Cavalry took to the field in 1867 with operations in Texas. Company K of the 9th was led by Captain William Frohock. In December 1867, Captain Frohock's company was attacked by several hundred Indians near the abandoned ruins of Fort Lancaster, Texas. In the three-hour battle, the Indians ran off the troopers' horses and tried to overrun the cavalrymen's positions. The cavalrymen, armed with the seven-shot Spencer carbine, turned back these attacks at a great loss to the Indians. The Indians broke off the attack and left the troopers with three men dead.

The main feature of the Model 1865 Spencer

Lieutenant Colonel George Crook. The 1872-1873 Tonto Basin War against the Indians was one of the most successful ever waged. For his triumph in this campaign, Crook was promoted to brigadier general. He is shown here in his 1880s campaign against the Apaches. *National Archives collection*

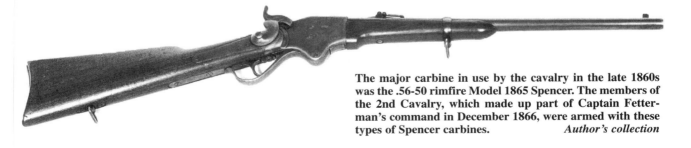

The major carbine in use by the cavalry in the late 1860s was the .56-50 rimfire Model 1865 Spencer. The members of the 2nd Cavalry, which made up part of Captain Fetterman's command in December 1866, were armed with these types of Spencer carbines. *Author's collection*

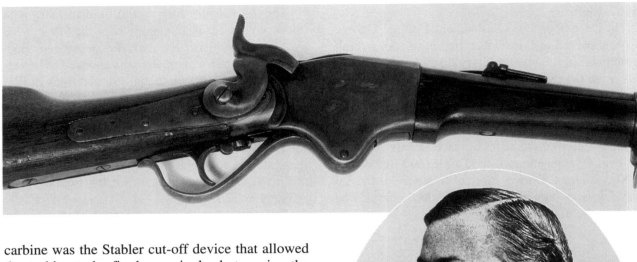

carbine was the Stabler cut-off device that allowed the carbine to be fired as a single-shot, saving the seven rounds in the magazine. The effectiveness of the Stabler system can be seen in a skirmish against the Comanches at Blanco Canyon, Texas, by members of the 4th Cavalry. In this engagement, Captain Robert Carter and five men were able to hold off the Comanches by using their carbines as single-shots until the last minute when they unlocked the Stabler cut-offs, unloaded their magazines and raced for the cover of the nearby arroyo. The Stabler device saved the troopers' lives.[11]

The year 1867 saw substantial field service against the Indians by the four new regiments of cavalry. In April, Captain Williams, 8th Cavalry, with 85 men out of Fort Whipple, Arizona successfully destroyed the Indian rancheria on the Verde and killed 50 Indians. On November 27, Custer's 7th attacked the Cheyenne village of Black Kettle in the Battle of Washita. Custer's attack on the village resulted in the destruction of a large portion of the Cheyenne winter food supplies. In the battle, Custer suffered 21 killed and 14 wounded. The 19th Kansas Vol. Cavalry, led by Colonel Crawford, was called out during Custer's 1867 campaign. The 19th Kansas was supplied with 500 Spencer carbines, while the settlers in Kansas were given Maynard carbines. The Spencer and Maynard carbines had been received as part of the State of Kansas' allotment of arms issued by the Ordnance Department. The 10th Cavalry, on September 25, broke the siege held by several hundred Indians at Beecher's Island. Major George Forsyth and 50 men armed with Spencer carbines had held Indians at bay for eight days until relieved by Captain Louis Carpenter of the 10th. As of June 1867, the Cavalry was armed as shown in the chart on page 92:

(above left) The Stabler cut-off device that was on a large number of the Model 1865 Spencers, allowed the carbine to be used as a single-shot arm and the seven-shot magazine to be kept in reserve. *Author's collection*

(above right) One of the most successful cavalry commanders in the West was Colonel Ranald Mackenzie of the 4th Cavalry. During most of the late 1860s and early 1870s, the 4th was armed with the Model 1865 Spencer carbine. *National Archives collection*

(facing page - top) Many of the Army's Indian scouts during this period were armed with Spencer carbines. *National Archives collection*

(facing page - bottom) At Beecher's Island, 50 men armed with Spencer carbines held out for eight days until relieved by a detachment of the 10th Cavalry. Lieutenant Beecher of the 3rd Infantry was killed in this fight. Lieutenant Beecher is shown, on the left, with a Lieutenant Cox. *The Herb Peck Jr. collection*

Carbines in the U.S. Cavalry[12]
June 30, 1867

	Sharps	Spencer .52 cal.	Spencer .50 cal.	Maynard
1st U.S.	683			84
2nd U.S.		189	853	
3rd U.S.	950	8	37	104
4th U.S.		307	678	
5th U.S.	930			
6th U.S.		68	745	
7th U.S.		119	374	
8th U.S.			620	
9th U.S.			620	
10th U.S.		80	409	
Total	**2,563**	**771**	**4,336**	**188**

During this period, the cavalry were not the only ones issued the late-war carbines. In March 1867, Major Frank North and Captain Luther North led a battalion of Pawnee scouts armed with Spencer carbines against the Sioux who were trying to block the passage of the Union Pacific Railroad. The Union Pacific Railroad workers were obtaining Joslyn and Ballard carbines and Springfield .58 caliber rifle muskets from the government for their protection against the Sioux. On September 20, 1867, Captain Barringer, Chief of the Commissary Department, was issued six Model 1865 Spencer carbines plus 250 rounds of ammunition. The Spencers were intended to protect the commissary's herd of cattle and remained in service for three years. At the end of three years, they were returned in unserviceable condition.[13]

The states received Spencer carbines from the Ordnance Department during October 1868. Colorado Territory was issued 500 Spencers and Nebraska 81. Captain Robinson, Company A, First Battalion, 2nd Brigade, California National Guard, requested on September 22, 1869 that his cavalry company be issued 80 Spencer carbines, and a like quantity of carbine slings, swivels and carbine cartridge boxes. The request had to be turned down since the state had already exceeded its quota of arms to equip the state militia.[14]

Warren Fisher Jr., treasurer of the Fogarty Rifle Co., wrote the Secretary of War on January 20, 1869, requesting a contract for either the Spencer or Fogarty repeating arms. Fisher states that the Boston factory had the capacity to manufacture 150 carbines or 120 rifles per day. Fisher's letter was forwarded to the Ordnance Department for their consideration. They responded that currently there were 33,000 serviceable caliber .50 Spencer carbines and 7,000 caliber .52 Spencer carbines in storage at the arsenals. The quantity of Spencers on hand would supply the needs of the service for several years and, therefore, no order should be given at this time.

The number of Spencers on hand was determined by a January 9 Ordnance Department directive instructing the arsenals to report the quantity of Spencer carbines in their inventories. The results are shown below:

Serviceable Spencer Carbines[15]
In Storage
January 1869

Location	cal. .52	cal. .50 w/Stabler	cal. .50 w/o Stabler
Allegheny	3,017	739	
Baton Rouge		258	50
Benicia		840	
Columbus	43	9,210	
Frankford	938	9,050	409
Fort Union		285	
Indianapolis	180		
Kenneche		3,000	
Leavenworth		2,304	8
St. Louis	160		16
San Antonio	430	135	30
Springfield		5,000*	
Vancouver		953	
Washington	1,598		28
Watervliet	428		738
Total	**6,794**	**31,774**	**1,279**

* This total is an estimate.

In addition to the above Spencers in storage, it is believed that all ten regiments of cavalry was armed with Spencers at this time. The total quantity of Spencer carbines in the federal government inventory, including both those in the field and those in storage, came to about 50,000 arms.

In January, the Ordnance Department also decid-

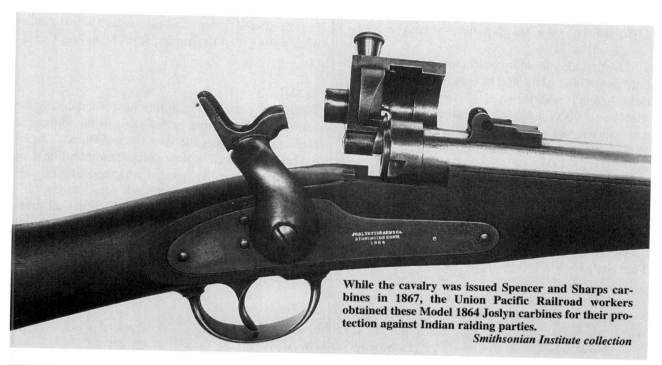

While the cavalry was issued Spencer and Sharps carbines in 1867, the Union Pacific Railroad workers obtained these Model 1864 Joslyn carbines for their protection against Indian raiding parties.
Smithsonian Institute collection

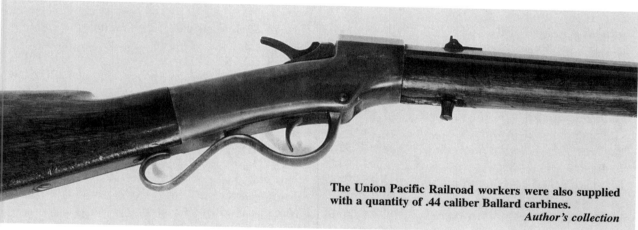

The Union Pacific Railroad workers were also supplied with a quantity of .44 caliber Ballard carbines.
Author's collection

ed to reduce the inventory at the St. Louis Arsenal by sending all serviceable carbines to Colonel Crispin at the New York Arsenal for resale. The following is the list of carbines sent to New York:[16]

Ballard	143
Burnside	7,136
Cosmopolitan	3,580
Gallager	6,360
Gibbs	114
Hall	1,747
Joslyn	1,613
Maynard	2,248
Merrill	2,495
Sharps & Hankins	182
Smith	5,195
Starr	1,860

In the years immediately after the war, the attention of the Ordnance Department turned to converting the small arms in storage to the new military center-fire .50-70 cartridge. On November 2, 1867, the Sharps Rifle Mfg. Co. entered into an agreement to alter all the percussion Sharps in the government arsenals to the centerfire .50-70 cartridge. All carbines with barrel diameters of over .5225 inches were sent to the Springfield Armory to have a barrel liner inserted. The Armory also replaced over 18,000 Sharps buttstocks. By October 1869, the Sharps factory had altered 31,098 carbines and

1,086 rifles for the government at a cost of $4.50 per alteration.[17]

The first altered Sharps carbines were sent to the field in October of 1868. On October 6, the Springfield Armory was directed to send 780 Spencer rifles and 1,000 altered Sharps carbines to the Leavenworth Arsenal for issue to the troops. In January 1869, the Chief of Ordnance for the Department of the Platte issued the following ordnance:

Fort McPherson:
500 altered Sharps carbines & 100,800 cartridges
Fort D.A. Russell:
600 altered Sharps carbines & 120,000 cartridges
Fort Sedgwick:
100 altered Sharps carbines & 20,000 cartridges

The altered Sharps carbines issued to Fort Sedgwick were given to Captain John Mix, 2nd Cavalry.[18]

1870-1873

By 1870, most ammunition purchased during the war was obsolete and of little use to the Ordnance Department. During the year, the Indianapolis Arsenal broke down all of their carbine ammunition and, as of November, the Arsenal was reporting no carbine ammunition in inventory. On October 27, 1870, the Ordnance Department directed all arsenals to report the quantity of small arms ammunition in storage. The chart below reflects the quantities of carbine ammunition on hand at the end of October, which totaled over 60 million rounds.

Serviceable Carbine Ammunition[19]
In Storage at U.S. Arsenals
October 1870

Ballard cal. .42 — *Copper* 2,110,132	Sharps & Hankins cal. .52 — *Metallic* . 755,063
Burnside cal. .54 — *Metallic* 6,789,267	Sharps cal. .52 — *Linen* 9,523,549
Colt cal. .44 — *Paper* 199,400	Sharps cal. .52 — *Paper* 891,926
Cosmopolitan cal. .50 — *Paper* 216,750	Sharps cal. .52 — *Skin* 200,750
Cosmopolitan cal. .52 — *Linen* 260,610	Smith cal. .50 — *Foil* 6,283,540
Gallager cal. .52 — *Foil* 1,525,485	Smith cal. .50 — *Rubber* 1,942,910
Gallager cal. .52 — *Metallic* 2,950,200	Spencer cal. .52 — *Metallic* 10,506,657
Hall cal. .52 — *Round Ball* 42,920	Spencer cal. .50 — *Metallic* 4,087,966
Joslyn cal. .54 — *Skin* 12,016	Starr cal. .54 — *Linen* 770,574
Maynard cal. .50 — *Metallic* 5,453,994	Starr cal. .52 — *Linen* 4,000
Merrill cal. .54 — *Paper* 1,149,522	Pistol Carbine cal. .58 — *Paper* 39,000
Musketoon cal. .69 — *Various* 449,630	Warner cal. .50 — *Copper* 927,372
Remington cal. .42 — *Metallic* 4,239,800	Wesson cal. .44 — *Metallic* 37,800

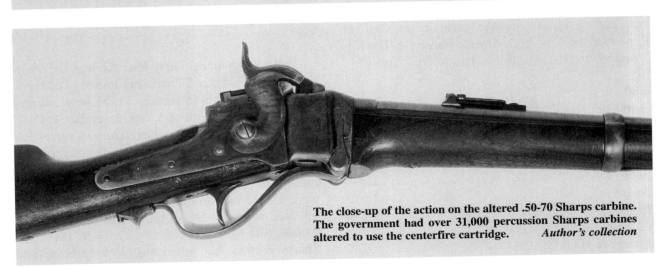

The close-up of the action on the altered .50-70 Sharps carbine. The government had over 31,000 percussion Sharps carbines altered to use the centerfire cartridge. *Author's collection*

The December 1870 reports from the field showed that the 9,700 carbines that were in cavalry use were almost equally divided between the altered Sharps carbine in .50-70 and the Model 1865 Spencer carbine in the .56-50 rimfire cartridge.

Carbines in the U.S. Cavalry[20]
December 1870

Regiment	M1868 Sharps	M1865 Spencers	Maynard
1st U.S. Cavalry	17	1,035	
2nd U.S. Cavalry	957		
3rd U.S. Cavalry	864	189	2
4th U.S. Cavalry		921	
5th U.S. Cavalry	992		
6th U.S. Cavalry		697	
7th U.S. Cavalry	963	4	
8th U.S. Cavalry	1,010	257	
9th U.S. Cavalry		857	
10th U.S. Cavalry	95	862	
Total	**4,898**	**4,822**	**2**

Altered Sharps carbines continued to be issued in greater numbers than the Spencer. A year and a half later, the number of Sharps had increased to over 7,100 while the quantity of Spencers had dropped to 2,200. Twelve months later, in June 1873, the number of Spencers had declined to only 162. The last Spencer reported in the cavalry was in March 1875, when the 4th Cavalry had 16 Spencer carbines on hand. The carbine of the U.S. Cavalry prior to the delivery of the M1873 Springfield was the M1868 altered Sharps. The decline in the quantity of Spencers and the increase in Sharps can be seen in the following chart:

Spencers and Sharps Carbines[21]
In Field Service
1871 - 1873

	M1865 Spencer			M1868 Sharps		
	9/71	6/72	6/73	9/71	6/72	6/73
1st Cavalry	363	352	26	557	673	789
2nd Cavalry				766	852	853
3rd Cavalry	72			926	907	872
4th Cavalry	1,003	682	52		640	838
5th Cavalry	1			850	771	805
6th Cavalry	924	576	74		144	870
7th Cavalry	4	4	4	780	767	692
8th Cavalry	91	10	1	1,020	890	858
9th Cavalry	867	618	5		638	667
10th Cavalry	73			845	853	886
Total	**3,398**	**2,242**	**162**	**5,744**	**7,135**	**8,130**

Pickets of the 1st U.S. Cavalry watching for Captain Jack's warriors in the rugged lava beds of Northern California during the Modac War.
National Archives collection

The major Indian campaigns during the later part of 1872 and early 1873 occurred in Northern California in the Modac War and in Arizona in Crook's Tonto Basin Campaign against the Apaches. Early on the morning of November 29, 1872, Troop B of the 1st Cavalry arrived at the Modac village of Indian leader Captain Jack to direct him and his followers to return to the reservation. Within minutes, shots were exchanged by both sides; the Modac War had begun. Captain Jack took his people to their stronghold in the lava beds. Over the next several months, the Modacs were able to hold the army at bay from their defensive position.

By early April 1873, the government had started talks to end the hostilities. On Good Friday, April 11, 1873, Brigadier General Edward R.S. Canby and two peace commissioners were killed by Captain Jack and his men while conducting peace talks. Shortly after Canby's death, the Modacs left the lava beds. On June 3, Captain Jack was captured and tried for Canby's death. He was convicted and hung at Fort Klamath on October 3, 1873.

In the one major engagement of the Modac War,

Post-War Period, 1866-1873

on January 17, 1873, the 225 Regulars of the 1st Cavalry and the 21st Infantry plus 100 militia were easily repulsed in their attacks on the lava beds. In this all-day action, the army suffered losses of nine killed and twenty-eight wounded, while the Modacs suffered no casualties. In one other skirmish, on April 26, an army reconnaissance force of 64 troopers was ambushed and suffered 25 killed and 16 wounded.

If the Modac War was a major disappointment for the Army, Crook's 1872-1873 campaign against the Apaches and Yavapais in Arizona was one of the most successful ever mounted against the Indians. In 20 engagements, between November 1872 and March 1873, the Army killed nearly 200 Indians. The campaign was conducted by nine troops made up of elements of the 1st and 5th Cavalrys. The winter campaign waged by Crook kept the Indians totally off guard. At the Battle of Skull Cave on December 28, the cavalry trapped and killed 76 out of a band of 100 Yavapais. By April 1873, the Indians were drifting back to the agencies. The Tonto Basin War was over. For several years, an uneasy peace prevailed over Arizona. For his success in this campaign, George Crook was promoted from lieutenant colonel to brigadier general.

Experimental Model 1870 Carbines

In August 1870, Springfield Armory was directed to fabricate 1,000 rifle muskets and 300 carbines each of the Springfield, Remington and Sharps breechloading design. Six months later, the Ward-Burton design was also added for fabrication. These four arms were issued for field trials to determine the design best suited for military use.

On May 2, 1871, Springfield Arsenal was directed to issue the experimental carbines to these locations:[22]

Augusta Arsenal	56 of each type =	168
Benicia Arsenal	56 of each type =	168
San Antonio Arsenal	56 of each type =	168
Omaha Depot	56 of each type =	168
Leavenworth Arsenal	84 of each type =	252
Total	**308**	**924**

The Springfield, Remington and Sharps carbines were sent from the arsenal on May 20. Companies C and E of the 2nd Cavalry and F and K of the 7th were the first units to receive these experimental arms in June. The first of the Ward-Burtons did not reach the field until a year later.

The Model 1870 Experimental carbines were in the following cavalry regiments: *(See table on next page).*

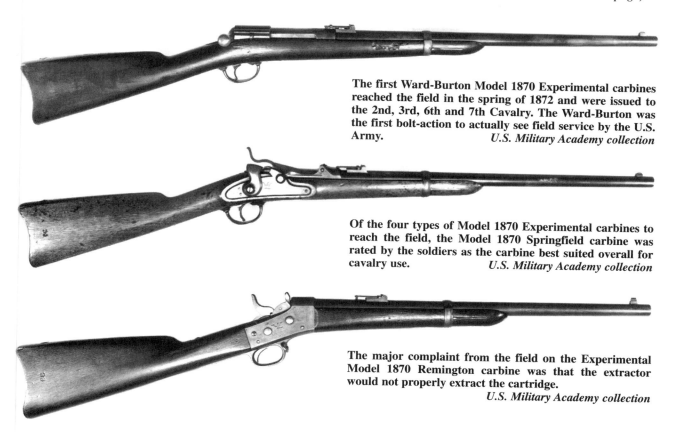

The first Ward-Burton Model 1870 Experimental carbines reached the field in the spring of 1872 and were issued to the 2nd, 3rd, 6th and 7th Cavalry. The Ward-Burton was the first bolt-action to actually see field service by the U.S. Army. *U.S. Military Academy collection*

Of the four types of Model 1870 Experimental carbines to reach the field, the Model 1870 Springfield carbine was rated by the soldiers as the carbine best suited overall for cavalry use. *U.S. Military Academy collection*

The major complaint from the field on the Experimental Model 1870 Remington carbine was that the extractor would not properly extract the cartridge. *U.S. Military Academy collection*

	Springfield		Remington		Sharps		Ward-Burton	
	6/72	6/73	6/72	6/73	6/72	6/73	6/72	6/73
1st Cavalry	48	17	51	29	49	57		33
2nd Cavalry	41	23	35	17	43	32	42	34
3rd Cavalry	7	10	7	11	7	7	14	13
4th Cavalry	27	15	20	14	26	54		23
5th Cavalry	4	10		3			54	17
6th Cavalry	44		44		42	18	56	30
7th Cavalry	53	32	56	27	56	51		30
8th Cavalry	28	18	25	17	28	22		25
9th Cavalry	27	20	27	20	27	20		24
10th Cavalry	NO EXPERIMENTAL CARBINES IN INVENTORY							
Total	**279**	**145**	**265**	**138**	**278**	**261**	**166**	**199**

M1870 Experimental Carbines (cal. .50) In Field Service[23] 1872 - 1873

Custer's Seventh Cavalry at the time of the 1874 Black Hills Expedition was listing in their regimental inventories of small arms both the .50-70 Sharps and .45-70 Springfield carbine. Shown here is Custer's wagon train during the Black Hills Expedition.
Little Bighorn Battlefield National Monument collection

The .50-70 centerfire Sharps carbine was the standard arm of the U.S. Cavalry at the time of the introduction of the .45-70 Springfield Model 1873 carbine into the field in 1874.
Author's collection

The most accepted experimental carbine in the field was the Springfield Trapdoor design. Captain Wells, Co. E, 2nd Cavalry, reported in September 1872 the typical view held on these carbines. As of September, Company E had in inventory 14 Remingtons, 16 Springfields, 18 Sharps and 21 Ward-Burton carbines. Captain Wells writes:

I have used the experimental arms a long time, and find that the Sharps carbine fouls in the breech by use, and after being fired a number of times, the lever bar does not work smoothly, also the extractor does not extract the shell properly. The Remington does not also extract the shell properly. The extractor only draws the shell a short distance, when they have to be taken out with the fingers.

Sometimes they expand, and can not be withdrawn without considerable difficulty. The Ward-Burton fails often to explode the cartridge, which is caused by the firing pin spring becoming weak. The extractor is good but as the piece is always cocked, and there are so many motions which consume time in working the gun, it is not in my opinion adopted for the cavalry service.

The Springfield has a good extractor. The firing pin does not become weak, and it is no more liable to get out of order than either of the other guns. It is well made and is in my opinion, the best carbine in use for cavalry service.[24]

The experimental carbines remained in the Cavalry service until September 1875 when the last Ward-Burton, Remington and Springfield carbines were reported in the 7th and 9th Cavalry.[25]

An ordnance board under the direction of Brigadier General Alfred Terry was established at New York City in September 1872 to select a breechloader system for military service. The board tested the arms already in the field as well as several other designs. In their report dated 3 May 1873, the board, like the soldiers in the field, found in favor of the Springfield Trapdoor system. The cartridge adopted for the Springfield was the .45/70. With the adoption of the Springfield system, the lives of the old Civil War Sharps and Spencer carbines in field service were nearing an end.

Footnotes — Post-War Period, 1866-1873

[1] NARG 156-110 and 108-77
[2] NARG 156-110
[3] NARG 156-21 Box 249
[4] NARG 156-21 Boxes 247-253
[5] NARG 156-21 and Roy Marcot, "Springfield Armory Conversions and Repairs to Spencer Repeating Carbines," *The Gun Report,* July 1980, pp. 65-67
[6] James A. Sawicki, *Cavalry Regiments of the U.S. Army,* Dumfires, VA, 1985, pp. 164-171; Frederick P. Todd, *American Military Equipage 1851-1872*
[7] NARG 156-21 Box 248
[8] NARG 156-21 Box 250
[9] *Ibid.*
[10] Robert M. Utley, Frontier Regulars, University of Nebraska Press, 1973, pp. 103-10
[11] Wayne R. Austerman, "Seven Rounds Rimfire," *Man at Arms,* March/April 1985, pp. 20-29
[12] NARG 156-110, Quarterly Report June 30, 1867
[13] NARG 156-21 Boxes 273 and 279
[14] Roy M. Marcot, *Spencer Repeating Firearms,* Irvine, CA, 1983, p. 129 and NARG 156-21 Box 276
[15] NARG 156-21 Box 269-274
[16] *Ibid.,* Box 282
[17] NARG 156-6, 1345 and 1365
[18] NARG 156-21 Box 273
[19] *Ibid.,* Boxes 277-283
[20] NARG 156-110 Report of December 31, 1870
[21] NARG 156-110 and 112
[22] NARG 156-6
[23] NARG 156-112
[24] Report of the Secretary of War Vol. 3, 1873, p. 157
[25] NARG 156-112 Report of September 30, 1875

A trooper, probably from Company A, 3rd Cavalry, armed with a Model 1873 Springfield Trapdoor carbine. He also is shown equipped with an earlier style Civil War era carbine sling. *The Herb Peck Jr. collection*

TRAPDOOR ERA
◆ 1874-1895 ◆

On the 28th of May 1873, Springfield Armory was notified to start production on the new Springfield carbine. It took several months to tool up for production. Between November 15 and the end of the year, 1,940 carbines and two rifles were fabricated at the arsenal.[1]

In early 1874, the San Antonio Arsenal requested that the new .45-70 Model 1873 carbines be shipped there for the spring campaign against the Comanches.[2] At end of March 1874, over 1,000 carbines were reported in field service in the following cavalry regiments:

**.45-70 Springfield Carbines[3]
In Field Service
March 31, 1874**

2nd Cavalry 168 carbines in Cos. D and I
4th Cavalry 80 carbines in Co. I
6th Cavalry 84 carbines in Co. E
10th Cavalry 670 carbines in Cos. B, C, E, H, I, K, L, M

Total   1,002

By September, the total number of Springfields in field service had increased to over 6,200 carbines — all cavalry regiments having been partially rearmed except for the 5th and 8th Cavalry regiments.

**Carbines in U.S. Cavalry[4]
September 1874**

	M1865 Spencer	M1868 Sharps	M1873 Springfield	M1870 Experimental Carbines
1st U.S. Cavalry	83	403	566	147
2nd U.S. Cavalry		118	975	2
3rd U.S. Cavalry		304	976	
4th U.S. Cavalry	23	5	937	
5th U.S. Cavalry	2	839		
6th U.S. Cavalry	4	6		11
7th U.S. Cavalry		859	926	7
8th U.S. Cavalry	1	833	751	43
9th U.S. Cavalry		610	209	39
10th U.S. Cavalry		2	952	
Total	**113**	**3,979**	**6,292**	**249**

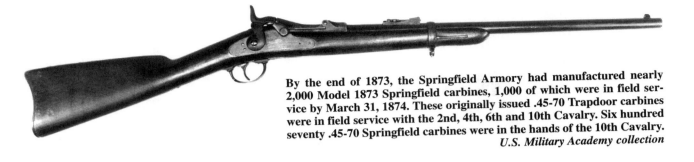

By the end of 1873, the Springfield Armory had manufactured nearly 2,000 Model 1873 Springfield carbines, 1,000 of which were in field service by March 31, 1874. These originally issued .45-70 Trapdoor carbines were in field service with the 2nd, 4th, 6th and 10th Cavalry. Six hundred seventy .45-70 Springfield carbines were in the hands of the 10th Cavalry. *U.S. Military Academy collection*

Location of Cavalry

1st U.S.	Washington Territory & California
2nd U.S.	Fort Laramie & Fort Ellis, Montana
3rd U.S.	Nebraska & Wyoming
4th U.S.	Texas
5th U.S.	Arizona
6th U.S.	Kansas
7th U.S.	Dakota Territory
8th U.S.	New Mexico
9th U.S.	Texas
10th U.S.	Fort Sill & Texas

One of the first engagements fought with the new Springfield carbine occurred on Saturday, August 22, 1874, at the Wichita Agency located thirty-seven miles south of Fort Sill. Colonel John Davidson and four troops of the 10th Cavalry arrived at the agency to disarm the Indians. When Davidson directed the Comanches to lay down their arms and surrender as prisoners of war, a full-scale battle erupted. In the two-day battle, the soldiers suffered three wounded and the hostiles one dead. After the smoke cleared, many of the Indians left the reservation but soon returned.[5]

The first Indian campaign in which the cavalry was armed with the .45-70 Springfield carbine was the Red River Campaign of August 1874 through April 1875. In this campaign, over two dozen engagements were fought between the cavalry and the southern plains Indians. General Sheridan's strategy of total warfare so wore down the Indians with constant threats of surprise attacks, bad weather and the lack of food that they returned to the reservation by April of 1875. One of the most successful cavalry skirmishes of the campaign occurred on the morning of September 28, 1874. In this action, Colonel Mackenzie with eight troops of the 4th Cavalry consisting of 21 officers and 450 enlisted men attacked an Indian encampment of several hundred lodges in the Palo Duro Canyon. The Indians fled before the cavalry with few casualties on either side. The cavalry captured and burned all their lodges and food stores and captured the Indians' entire pony herd of 1,424 head. Mackenzie kept the best of the herd and had over a thousand ponies destroyed.[6]

In 1875, all the cavalry regiments had been refitted with Springfield carbines as shown in the following chart:

Carbines in the Cavalry[7]
September 1875

	Sharps	M1873 Springfield	M1870 Experimental
1st Cavalry	3	834	
2nd Cavalry	10	949	
3rd Cavalry	16	850	
4th Cavalry	1	920	
5th Cavalry	193	530	
6th Cavalry		518	
7th Cavalry	23	808	55
8th Cavalry	24	668	
9th Cavalry	24	615	7
10th Cavalry	10	857	
Total	**304**	**7,549**	**62**

LITTLE BIGHORN
June 25-26, 1876

At noon on Sunday, June 25, 1876, Custer and his Seventh Cavalry rode into the valley of the Little Bighorn and into history. Custer divided his command into three columns. Captain Frederick Benteen with 125 men from Troops D, H and K

Lieutenant Colonel George Armstrong Custer of the 7th Cavalry. Photo was taken shortly before his death at the Little Bighorn on June 25, 1876.
Little Bighorn Battlefield National Monument

The only known survivor of Custer's five troops of cavalry was the horse on the right, Comanche, shown here with its handler Gustave Korn at Fort Meade, South Dakota, in 1881. Korn was killed at Wounded Knee in 1890. The soldier on the right is Charles S. Isley. *Little Bighorn Battlefield National Monument*

were to scout to the south. Major Marcus Reno, with 112 men from Troops A, G and M, was directed to attack the upper end of the Indian village with the understanding that if help was needed Custer would come to their support. At about 3 p.m., Reno attempted to enter the village but pulled up short of it. Here he established a skirmish line with 90 troopers. The balance of Reno's command was detailed to hold the horses. The troopers quickly ran low on ammunition. Every second man was ordered back to the saddlebags for more ammunition. The carbines overheated and the cartridges became difficult to eject. At this point, Reno ordered his men to take shelter in a stand of trees. Shortly thereafter, he ordered a retreat across the river to the bluffs. In this action, which had lasted about 45 minutes, Reno's command had suffered 3 officers and 29 enlisted men killed, 18 missing, and 10 or 11 wounded.[8] After Reno reached the bluffs, Captain Benteen came to his assistance.

After leaving Reno, Custer had continued north along the bluffs. The Indian encampment of over 7,000 Sioux and Cheyenne lay hidden from Custer's view. The lead elements of the column attempted to cross the river into the Indian camp

Here is the burial site of Captain Myles Keogh, Troop I commander and eight troopers from F Troop. Keogh's grave marker is reflecting his wartime rank of colonel. Note that when this photo was taken, the horse skeletons were still in view. *U.S. Military History Institute, MOLLUS*

but were turned away. Dismounting and forming skirmish lines, 1st Lieutenant Calhoun's Troop L and Captain Myles Keogh's Troop I made up the North skirmish line, while Captain Tom Custer's Troop C, Captain Yates' Troop F and 1st Lieutenant A.E. Smith's E Troop made up the South skirmish line. The cavalrymen were armed with 100 rounds of .45-70 ammunition plus 24 rounds for their Colt revolvers.[9]

The 1,500 Indians engaged in the Custer fight concentrated their fire on the skirmish lines and were able to overrun these positions. Custer and several of his men made their stand on what is now known as Last Stand Hill. As the fire slackened off, the Indians overwhelmed the few survivors in hand-to-hand combat. There were 42 bodies and 39 dead horses found on this hill. In the two-day battle, the 7th Cavalry suffered 268 killed and 44 wounded.[10] Included among the dead were Custer, his brothers Tom and Boston, and nephew Harry Armstrong Reed. The Indian losses are not known for sure but are believed to have been about 100 — many of whom died of their wounds.

After the battle, Sergeant Charles Windolph of Troop H under Benteen's command stated that he believed that fully half of the Indians that he saw were armed with rifles and 25 to 30 percent were carrying modern repeating rifles. The archeological study that took place at the Custer battle site after the 1983 grassfire confirmed that the Indians were, in fact, better armed and, indeed, out-gunned Custer's command. The archeological study concluded that as many as 192 repeating firearms were in use by the Indians. These repeating arms included Henrys, Model 1866 and 1873 Winchesters plus a few Spencers. In addition, they also carried .50-70 Sharps and Springfields. Among the percussion arms were a .54 caliber Starr and a Maynard carbine, .577 caliber Enfield rifles and other muzzle-loading rifles and shotguns.[11]

One of the problems raised by Major Reno in his July 11th report on the battle was the failure of the extractor to eject the spent cartridge from the chamber. Captain Otho E. Michaelis, Chief of Ordnance for the Military Department of Dakota, reported on September 29, 1876 that he had examined only one carbine from the battle that had a cartridge stuck in the action due to the head breaking off. He placed the blame for this problem on defective cartridges and the hard usage of the cartridges before being inserted into the chamber.[12]

Over a two-year period, between January 1876 and December 31, 1877, the Seventh Cavalry was issued the following quantities of ordnance and ordnance stores:

Ordnance and Ordnance Stores[13]
Issued to 7th Cavalry
January 1876 - December 31, 1877

Class VI
- 1,059 Springfield Carbines cal. .45
- 1,225 Revolvers cal. .45
- 125 Springfield Rifles cal. .45

Class VII
- 635 Saddles
- 2,140 Saddle blankets
- 462 Saddle bags, pairs
- 2,104 Horse brushes
- 1,095 Curry combs
- 1,288 Pistol holsters
- 650 Carbine slings
- 625 Carbine swivels
- 125 Carbine cartridge boxes
- 240 Carbine sockets and straps
- 1,272 Canteens
- 1,878 Haversacks
- 1,605 Tin cups
- 1,665 Knives, forks, and spoons
- 820 Cartridge belts

Class VIII
- 286,000 Carbine ball cartridges cal. .45
- 5,000 Carbine blank cartridges cal. .45
- 141,000 Revolver ball cartridges cal. .45
- 5,600 Revolver blank cartridges cal. .45
- 148,000 Rifle ball cartridges cal. .45

Miscellaneous
- 1,200 Entrenching tools and scabbards
- 1,180 Jointed ramrods for carbines
- 1,180 Extractors for headless cartridge shells

MAGAZINE ARMS
Hotchkiss Carbines

The only carbine other than the Springfield to be issued to the cavalry during the latter part of the 1870s and early 1880s was the five-shot tubular magazine-fed Hotchkiss carbine. The magazine, located in the butt of the stock, was loaded through a trap in the buttplate.

A year after the Custer defeat, Captain Michaelis, from his office in St. Paul, Minnesota, had requested that the cavalry should be issued magazine arms for trials.[14] Unknown to Michaelis, on November 21, 1877, Congress had authorized an ordnance board to select a magazine arm for military service. The board's report of September 23, 1878, found in favor of the Hotchkiss bolt action design with $20,000 appropriated for their procurement.

The Winchester Repeating Arms Company held the rights to the Hotchkiss design. In December 1878, Winchester agreed to supply the Springfield Armory with the Hotchkiss breech system, magazine, guard and buttplate complete, plus the stock furnished, turned and machined up to the portion of the lower barrel band, for $12 each. The armory supplied the proofed barrels, barrel bands, rod tips and swivels. The arms were inspected and assembled at the armory at a cost to the government of $6.00 apiece.[15] The total cost of $18 per arm allowed the government to place an order for up to 1,100 but they settled on 500 Hotchkiss carbines and a like number of rifles. The ordnance inspector for these arms was Captain Green.

By June 1879, the Hotchkiss arms were ready for issue. On June 26, the armory forwarded 176 Hotchkiss carbines to Fort Lincoln. Within the next two months, all 1,000 arms had been sent forward for issue.

Hotchkiss Arms Issued 1879[16]
From Springfield Armory

Fort Abraham Lincoln	400
San Antonio Arsenal	400
Vancouver Arsenal	100
Benicia Arsenal	100
Total	**1,000**

The Hotchkiss carbines were issued to the 1st, 4th, 5th, 8th and 10th Cavalry. It appears that two troops in each regiment were armed with the Hotchkiss carbine. In the 4th Cavalry, they were issued to Captain Wint's Troop L and 1st Lieutenant Alex Rodgers' Troop A.[17] The major problem with these carbines were the weak mainsprings, which had to be replaced, and some extractor breakage.

In the fall of 1879, the 4th and 5th cavalry was armed with both the Hotchkiss and Springfield carbines and were part of the force engaged in the Ute War in the Northwest. The following year, Grierson's 10th Cavalry was in the field against the Apache leader Victorio. Grierson placed his command along the Rio Grande River to block the Apaches' path when they re-entered Texas from Mexico. In several engagements with Victorio's Apaches, Grierson was able to force them back into Mexico where Victorio was killed along with 60 of his men in a battle with Mexican troops on October 15, 1880.

The Hotchkiss carbine, in use during this period, was the 1st Model. This model was equipped with a circular knob on the right side of the stock that controlled both the magazine cut-off and the safety function. In 1880, Winchester offered to replace the 1st Model Hotchkiss with the new improved design that replaced the magazine cut-off safety knob on the side of the stock with levers on either side of the receiver. This new improved model is known as the 2nd Model Hotchkiss.

The Ordnance Department, in January 1881, directed the troops in the field to turn in their Hotchkiss carbines and rifles so that the alterations could be made by the Winchester factory. The new stocks were to be provided by Springfield. In February, both the Benicia Arsenal and the ordnance depot at Fort Abraham Lincoln forwarded their Hotchkiss arms to Springfield Armory.[18] Springfield, after receiving the arms, forwarded them to Winchester for alteration.

As the altered arms (2nd Model) were returned from the factory, they were issued to Mackenzie's 4th Cavalry. By the middle of May, 414 of the carbines had been converted by Winchester. The first 210 Second Model Hotchkiss carbines were issued to the 4th in April.

2nd Model Hotchkiss Carbines[19]
Issued to 4th Cavalry
April 1881

Troop H, Fort Reno, I.T.	60 Hotchkiss Carbines
Troop G, Fort Reno, I.T.	50 Hotchkiss Carbines
Troop M, Fort Canadian, I.T.	60 Hotchkiss Carbines
Troop C, Fort Sill, I.T.	40 Hotchkiss Carbines

By July, the 4th Cavalry had 620 Hotchkiss carbines on hand. This quantity of arms would furnish the entire regiment. The 4th was the only cavalry regiment to have been issued the 2nd Model

The 6th Cavalry at Fort Bayard, New Mexico, about 1880, in a training exercise to condition their horses to carbine fire.
National Archives collection

In 1879-1880, the 1st Model Hotchkiss carbines were issued to two troops of the 1st, 4th, 5th, 8th and 10th Cavalry. The main complaint from the field on the Hotchkiss carbine was the weak mainspring which had to be replaced.
Smithsonian Institute collection

Hotchkiss carbine. The 4th saw action during the 1880s in Arizona against the Apaches. At Horseshoe Canyon on April 23, 1882, five troops of the 4th attacked a band of Apaches. In the action, the 4th suffered five killed and seven wounded while the Apaches substained thirteen killed. After the skirmish, the Apaches crossed over into Mexico and the pursuit was taken up by other cavalry units.

Issues of Hotchkiss Carbines[20] by Fiscal Year

Fiscal Year Ending June 30, 1879 . . . 201, 1st Model
Fiscal Year Ending June 30, 1880 . . . 302, 1st Model
Fiscal Year Ending June 30, 1881 . . 490, 2nd Model
Fiscal Year Ending June 30, 1882 . . 112, 2nd Model

Records indicate that 572 Hotchkiss carbines were assembled at the armory; therefore, the balance of the carbines issued were converted rifles.[21]

The story of magazine arms in military field service during the 1880s came to a conclusion in 1885. In that year, the Ordnance Department sent to the field 715 Lee rifles and a like number of Chaffee-Reeses and Hotchkisses for field testing. These arms were issued to all branches of the services. In the cavalry, they were assigned to Troops A, B, I and K of the 2nd; Troops A, F and H of the 4th; Troop H of the 5th and Troops A, I and M of the 10th Cavalry.[22] When the results were tallied, the best of the three magazine rifles was the Lee. The surprise was that when compared to the Springfield rifle for overall use, the Springfield won hands down. The results were Lee 10, Chaffee-Reese 3, Hotchkiss 4 and Springfield 46. With this report in hand, on December 15, 1885, it would be an additional ten years before the cavalry would be armed with magazine carbines.[23]

APACHE WARS
1881-1886

Between the outbreak of hostility with the Apaches in Arizona (August 1881) and Geronimo's surrender to General Miles (September 4, 1886),

Springfield Armory sent forward for issue 7,926 Model 1877 Springfield carbines.

Model 1877 Springfield Carbines[24]
Issued during the Apache Wars
1881 - 1886

Fiscal Year Ending June 30, 1882	1,103
Fiscal Year Ending June 30, 1883	614
July 21, 1883	1
November 11, 1883	500
January 19, 1884	200
March 22, 1884	250
May 3, 1884	200
May 10, 1884	1,000
June 21, 1884	1
July 26, 1884	26
September 6, 1884	40
November 29, 1884	1,100
December 13, 1884	50
January 17, 1885	2,400
February 2, 1885	1
February 6, 1886	300
April 24, 1886	100
September 4, 1886	40
Total	**7,926**

On September 4, 1886, the first 200 Model 1884 Springfield carbines with the Buffington rear sights were sent to the field for issue. By the end of the year, an additional 4,200 Model 1884 carbines had been sent forward as well as 722 additional Model 1877 carbines.

The final phase of the Apache Wars broke out in May 1885 when 130 Apaches, including Geronimo, fled the reservation for Mexico. Despite over 3,000 troops deployed along the Mexican border, Geronimo was able to evade both Mexican and U.S. forces until January 13, 1886, when talks were opened with Lieutenant Marion Maus of the U.S. Cavalry. Geronimo agreed to meet with General Crook to discuss returning to the reservation. This meeting with General Crook took place on March 27. Geronimo agreed to surrender, but during the night had a change of heart and fled into the mountains with 21 men and 13 women. Because of his failure to bring in Geronimo, Crook asked to be relieved of his command, which was turned over to Brigadier General Nelson Miles.

In addition to the 2nd, 4th and 10th Cavalry, Miles deployed several infantry regiments to help prevent the Apaches from crossing and recrossing the Mexican border. Miles also established a special

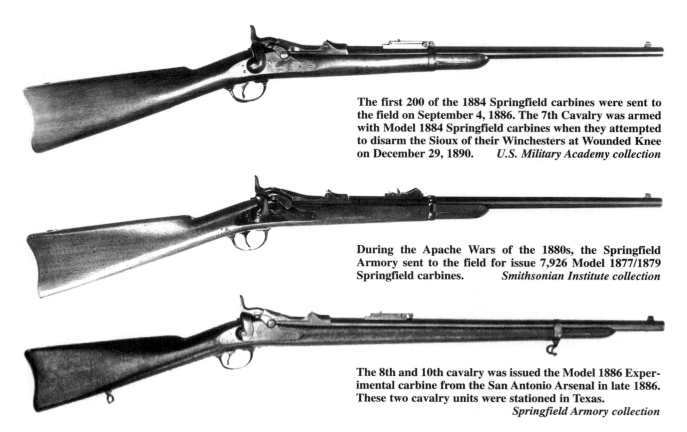

The first 200 of the 1884 Springfield carbines were sent to the field on September 4, 1886. The 7th Cavalry was armed with Model 1884 Springfield carbines when they attempted to disarm the Sioux of their Winchesters at Wounded Knee on December 29, 1890. *U.S. Military Academy collection*

During the Apache Wars of the 1880s, the Springfield Armory sent to the field for issue 7,926 Model 1877/1879 Springfield carbines. *Smithsonian Institute collection*

The 8th and 10th cavalry was issued the Model 1886 Experimental carbine from the San Antonio Arsenal in late 1886. These two cavalry units were stationed in Texas.
Springfield Armory collection

Trapdoor Era, 1874-1895

Photo of Army officers and Indian scouts taken during the Apache Wars in the 1880s. The officers are armed with both the Springfield carbine and rifle.
Arizona Historical Society Library collection

unit to track Geronimo wherever he went. The special unit was led by Captain Henry Lawton of Troop B, 4th Cavalry. Lawton's command consisted of 35 troopers from his own unit, 30 Apache scouts and 20 members of Company D, 8th Infantry. This force left Fort Huachuca on May 5, 1886.[25] Over the next four months, Lawton and his command traveled over 2,000 miles in pursuit of Geronimo and his band.

Finally, on August 25, Lieutenant Charles B. Gatewood, 6th Cavalry — the only officer in Lawton's command that Geronimo knew — entered the Apache village to discuss peace. The peace proposal sent by General Miles through Gatewood was to surrender and be sent with their families to Florida or stay out and fight to the end. When Geronimo learned that the reservation Apaches were already being sent to Florida, he agreed to surrender to Miles. The surrender occured at Skeleton Canyon, 65 miles southeast of Fort Bowie, on September 4. Geronimo and his people lived in the East until October 1894 when they were resettled at Fort Sill. Here he lived the remainder of his life, dying in 1909. With Geronimo's surrender to Miles, the Apache Wars had come to an end.[26]

Cavalrymen from the 10th Cavalry with their Springfield carbines at Fort Apache c.1890.
Arizona Historical Society Library collection

Model 1886 Experimental Carbine

During the latter part of the Apache Wars, the Ordnance Department had issued an experimental model carbine more in keeping with the longer standard rifle. For many years, the cavalry had been issued a few rifles. During the Nez Perce Uprising in 1877, the 7th Cavalry had been issued 10 rifles per company.

One of the major complaints against the Springfield carbine was its short range. Therefore, in 1884, the Board on Cavalry Equipment recommended that a carbine be manufactured with a 28-inch barrel. The carbine that was finally adopted was equipped with a 24-inch barrel with three grooves, a stock nearly the length of the arm, one barrel band, sling swivels, saddle bar and ring. These carbines, known as the Model 1886 Experimental carbine, were supplied with Buffington rear sights.

Springfield Armory was notified by letter on June 14, 1885, that for the upcoming fiscal year 1,000 experimental model carbines were to be manufactured at the armory. The new experimental carbines were to be equipped to use the rifle cartridge.[27] The weekly production reports from the armory indicate that 560 of the new experimental carbines were fabricated the week ending April 10, 1886, and the balance of the 1,000 carbines the following week. During this two-week period, these experimental carbines were the only arms manufactured at the armory.[28]

The new carbines remained in arsenal storage for nearly four months until Order for Supply #1202 was issued for all 1,000 carbines. The first 315 carbines were sent to the field on August 13 and the balance was forwarded on August 16 and 17. They were sent to the ordnance depots at Fort Lincoln and Cheyenne as well as the arsenals at Fort Leavenworth, San Antonio, Vancouver Barracks and Benicia. The San Antonio Arsenal issued 185 of the experimental carbines to the 8th and 10th Cavalry. The Model 1886 Experimental carbines arrived too late for the Apache Wars, but a few of the arms may still have been in use during the 1890 outbreak of the Ghost Dance.[29]

In addition to the arms in field service and in arsenal storage, the ordnance depots were listing over 1,100 Springfield carbines on hand as of January 31, 1890.

Springfield Carbines[30]
In Ordnance Depot Storage
January 31, 1890

	Carbines .50 cal.	Carbines M1884	Cartridges .45 cal.
Cheyenne	59		
Omaha		59	2,000
Fort Leavenworth		568	114,000
Fort Snelling		157	637,540
Vancouver Barracks		340	72,889
Total	**59**	**1,124**	**826,429**

Trapdoor Era, 1874-1895

Cavalrymen, probably from the 4th Cavalry, armed with Hotchkiss carbines. The Indian scouts are armed with Springfield rifles. The photo was taken in 1886, the last year of the Apache Wars.
Arizona Historical Society Library collection

Members of the 3rd Cavalry in 1890 at Fort Davis, Texas. Many of the cavalrymen are shown with their Springfield carbines.
U.S. Military History Institute

The 7th Cavalry at Pine Ridge in 1890 during the Wounded Knee Campaign.   *U.S. Army Military History Institute*

The 10th Cavalry in South Dakota in 1891.   *U.S. Military History Institute*

These same ordnance depots were also showing in inventory 992 M1884 Springfield rifles, 31 Officers Model Rifles and 10 Springfield shotguns.[31]

GHOST DANCE
1890-1891

By 1890, the Northern Indian reservations were in a state of great excitement over the Ghost Dance. The Indians believed that by praying, dancing and singing to the spirits, they would be reunited with their dead relatives and also rid themselves of the white man. The Sioux further believed that wearing ghost shirts would stop the white man's bullets.

Alarmed that this new religion could lead to hostilities, the Army deployed the 1st, 2nd, 5th, 6th, 7th, 8th and 9th Cavalry to keep the peace. On the morning of December 29, 1890, at Wounded Knee Creek, Colonel James W. Forsyth and 500 members of the 7th Cavalry were proceeding to disarm 120 Sioux of their Winchester rifles. In the process of turning over their rifles, shots were fired. When the smoke cleared, 25 soldiers were dead and 39 wounded. The Indians had suffered 146 casualties, of which 62 were women and children, plus an additional 50 wounded. In this action, the 7th was armed with the M1884 Springfield carbine. General Miles took up the task of returning peace to the region, which he was able to do when the last of the Sioux surrendered to him on January 15, 1891.

During the Indian Wars, 1866-1890, over 1,000 combat actions took place; the cavalry had been engaged in the majority of these encounters. Army reports indicate that 69 officers and 879 enlisted men were killed in action during this span of time.[32]

The Springfield carbine remained in the hands of the cavalry for the first half of the 1890s. During this period, over 7,700 carbines were sent to the field for issue.

Trapdoor Era, 1874-1895

Two cavalrymen (a corporal and a private) shown with their Model 1877/1879 Springfield carbines.

The Herb Peck Jr. collection

A cavalryman dressed in a later-period style uniform shown with his Trapdoor Springfield carbine.
The Herb Peck Jr. collection

Troop Issue
Springfield Carbines[33]
M1884

Fiscal Year Ending June 30, 1890	3,188 carbines
Fiscal Year Ending June 30, 1891	1,420 carbines
Fiscal Year Ending June 30, 1892	1,669 carbines
Fiscal Year Ending June 30, 1893	1,443 carbines
Total	**7,720 carbines**

While these carbines were being issued, the Ordnance Department was taking the first steps to replace the single-shot Trapdoor design with the adoption of the magazine-fed Krag-Jorgensen in 1892.

Footnotes — Trapdoor Era, 1874-1895

[1] Albert J. Frasca and Robert H. Hall, *The .45-70 Springfield,* Northridge, CA, 1980, pp. 6-7
[2] Louis A. Garavaglia and Charles G. Worman, *Firearms of the American West 1866-1894,* Albuquerque, NM, 1985, pp. 32-33
[3] NARG 156-112, Report of March 31, 1874
[4] *Ibid.,* Report of September 30, 1874
[5] Colonel W.S. Nye, *Carbine and Lance — The Story of Old Fort Sill,* Norman, 1937, pp. 206-212
[6] *Ibid.,* pp. 221-228, Robert M. Utley, *Frontier Regulars,* Lincoln, NB, 1973, pp. 219-233
[7] NARG 156-112, Report of September 30, 1875
[8] Wayne M. Sarf, *Great Campaigns: The Bighorn Campaign, March-September 1876,* Conshohocken, 1993, pp. 204-206, 221
[9] Robert P. Jordan, "Ghosts on the Little Bighorn," *National Geographic,* December 1986, pp. 791-793
[10] Wayne M. Sarf, op. cit., pp. 260 and 309
[11] Douglas D. Scott and Richard A. Fox Jr., *Archaeological Insight into the Custer Battle,* Norman, OK, 1987, p. 112; Wayne R. Austerman, "Maza Waken For the Sioux," *Man at Arms,* January/February 1988, pp. 24-25
[12] James S. Hutchins, "Captain Michaelis Reports on Army Weapons and Equipment on the Northern Plains, 1876-1879," *Man at Arms,* January/February 1988, pp. 30-31
[13] *Ibid.,* p. 34
[14] *Ibid.,* p. 31
[15] NARG 156-1354, dated December 14, 1878
[16] *Ibid.*
[17] NARG 156-1362
[18] *Springfield Research Service Serial Numbers of U.S. Martial Arms, Vol. 2,* Hotchkiss Rifles and carbines, pp. 151-156
[19] NARG 156-1354
[20] Secretary of War Reports in Vol. 3 of 1879, 1880, 1881, 1882
[21] NARG 156-1383
[22] Secretary of War Reports 1886, Vol. 3, p. 526, Appendix 35
[23] *Ibid.,* and Claud E. Fuller, *The Breech-Loader in the Service 1816-1917,* New Milford, CT, 1965, pp. 342-343
[24] Frasca and Hill, op. cit., p. 379 and NARG 156-1383
[25] Richard E. Killblane, "Arizona Tiger Hunt," *Wild West,* December 1993, pp. 42-49
[26] Robert M. Utley, op. cit., pp. 387-389
[27] Frasca and Hill, op. cit., pp. 151-152
[28] NARG 156-1383
[29] *Ibid.,* and 156-1354; Garavaglia and Worman, op. cit., p. 59
[30] NARG 156-105
[31] *Ibid.*
[32] Robert M. Utley, op. cit., p. 412
[33] Frasca and Hill, op. cit., p. 379

Roosevelt and his Rough Riders shown on Kettle Hill after its capture by them and the 10th Cavalry on July 1, 1898.
National Archives collection

Foreign Conflicts 1898-1905

A cavalryman armed with a Model 1896 Krag carbine and a Colt Single Action Army revolver. He is also shown equipped with a .30-40 cartridge belt.
The Herb Peck Jr. collection

U.S. MAGAZINE KRAG
◆ 1892-1905 ◆

By the late 1890s when U.S. involvement in foreign countries arose, the standard arm of the United States Cavalry was the Krag carbine — the first arm to be issued using the new smokeless powder .30-40 cartridge. The unique feature of this bolt action carbine is its breechloading five-shot magazine, the loading gate located on the right side of the receiver. The advantage of the Krag, as the Ordnance Board saw it, was its ability to be used as a single-shot arm while a cut-off lever held the magazine in reserve. The .30-40 cartridge has a 220 grain bullet and is charged with 36 grains of smokeless powder. The muzzle velocity of the cartridge is stated at 2,000 f.p.s.[1]

The inventors of the Krag were Captain Ole Krag, superintendent of the Kongberg Armory, and its chief armorer, Erik Jorgensen of Norway. In 1889, the two men were issued a patent on their bolt action design. In the same year, Denmark adopted it as their standard arm. Norway followed suit a few years later.

The Krag story in the United States started with the establishment of an Army Ordnance Board on November 24, 1890 by General Order No. 136 "to consider and recommend a suitable magazine system for rifles and carbines for the military service." The board gave a deadline of July 1892 for interested inventors to submit their arms for trial. In all, 53 arms were submitted for trials; of these, the No. 5 Krag-Jorgensen was found best adaptable for military use. The board's results of August 19, 1892, were forwarded to Acting Secretary of War L.A. Grant for his approval, which was given on September 15, 1892.[2] The American inventors, outraged by this development, petitioned Congress to direct that additional tests be conducted. Such tests were conducted in early 1893, but the results confirmed the previous year's verdict in favor of the Krag-Jorgensen.[3]

Production on the Krag started in late 1893 and by the fall of the following year, the first of the Krag rifles were ready for issue. The first unit to be issued the Krag was the 4th Infantry on October 6, 1894. By June 1895, all infantry regiments had been rearmed with the Krag while the cavalry was still armed with the .45-70 Trapdoor Springfield carbine.[4] The Krag carbine was approved for production on May 23, 1895. By the end of the year, 1,378 carbines had been manufactured, and by June 30, 1896, this total had risen to 7,111 carbines. The first of the Krag carbines were issued to the cavalry on March 10, 1896. The entire cavalry was rearmed with Krags by June 1896.[5]

Production on the Krags continued and by September 1897, the total number of Krag carbines in storage at Springfield Armory had risen to over

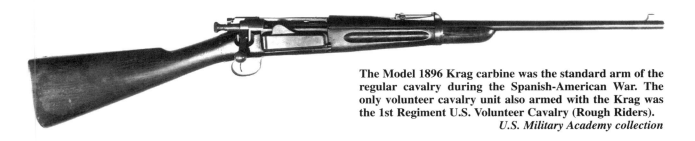

The Model 1896 Krag carbine was the standard arm of the regular cavalry during the Spanish-American War. The only volunteer cavalry unit also armed with the Krag was the 1st Regiment U.S. Volunteer Cavalry (Rough Riders).
U.S. Military Academy collection

By the outbreak of the Spanish-American War, Nelson Miles was commanding general of the Army. During the war, he directed the recruitment and training of the American forces. He was also the commanding officer of the troops that occupied Puerto Rico.

National Archives collection

11,000. This quantity of arms had grown to nearly 15,000 by the outbreak of the Spanish-American War in April of 1898. The following chart shows the carbine inventory at the Armory during this period.

Carbines on Hand[6]
Springfield Armory

	Sept. 30, 1897	Mar. 30, 1898
Sharps	8,225	0
Spencer	1,982	0
Springfield M1873	581	581
Springfield M1884	3,796	3,787
Krag M1895	2	2
Krag M1896	10,995	14,813

As of March 1898, Springfield Armory was also reporting over 51,000 Krag rifles and 110,000 Trapdoor rifles in storage.

Spanish-American War

At the outbreak of the war on 20 April 1898, the Regular Army consisted of 28,183 officers and men, of which less than 6,000 were cavalry. Except for a portion of the 3rd and 6th Cavalry, which were in the East, the remainder of the cavalry regiments were stationed at 31 posts in the West. For deployment to Cuba, the cavalry was directed to proceed to Chickamauga Park, Georgia. The last troop from Fort Sill left for the East at 2 p.m. on April 18. These men from Troop E, 1st Cavalry were commanded by Captain W.C. Brown.[7]

In addition to the Regulars, President William McKinley called for 200,000 volunteers; most of this total represented infantry units. The states of Illinois, Texas and Ohio supplied a cavalry regiment. Nine cavalry troops were raised from the states of Kentucky, Pennsylvania, New York, Nevada and Utah. Only the troops from Pennsylvania, New York and Utah were to deploy outside of the United States. The three Philadelphia cavalry troops that were deployed to Puerto Rico consisted of the Sheridan, Governor's and Philadelphia City Troops. They had been mustered into federal service between May 7 and 13. The Pennsylvania troops were armed with the .45-70 Trapdoor carbines of which the state had reported 360 in storage. Half of the 360 carbines were issued to the troops mustered into federal service.[8]

While the volunteers were being issued the .45-70 Trapdoor, one unit (the 1st Regiment, U.S. Volunteer Cavalry) was to be totally armed with the new Krag Model 1896 Carbine. This unit was popularly known as the Rough Riders. The colonel of the regiment was Leonard Wood and its lieutenant colonel, Theodore Roosevelt. During the Geronimo campaigns of the 1880s, Wood had been awarded the Medal of Honor. At the outbreak of the war, Roosevelt had been the Assistant Secretary of the Navy.

The Rough Riders had been organized at San Antonio, Texas, in early May. Three troops of the regiment were from Arizona, one from the Oklahoma Territory and New York, five from New Mexico and two from the Indian Territory.[9] The regiment consisted of 47 officers and 994 enlisted men.

During their stay at San Antonio, the Rough Riders had been issued 5½-inch barrel Colt Single Action revolvers and the Model 1896 Krag carbines. The Krag carbines that were issued to the regiment had been obtained through the influence

Lieutenant Colonel Theodore Roosevelt, 1st Regiment U.S. Volunteer Cavalry (Rough Riders). Roosevelt had been Assistant Secretary of the Navy prior to the outbreak of the war.
National Archives collection

of their commanding officers. The Krags issued to the Rough Riders are found in the serial ranges of 27,000, 28,000, 68,000, 74,000 and 77,000.[10] They had been shipped from the Springfield Armory since the beginning of the war. As of April 1, the San Antonio Arsenal had been reporting only 39 Krag Carbines and 82 Krag Model 1892 rifles in storage.[11]

Tampa, Florida, was selected as the jumping off point for the invasion of Cuba. In command of the Fifth Corps at Tampa was Major General William Shafter (1835-1906). Shafter had been an officer during the Civil War when he had been awarded the Medal of Honor. After the war, he remained in the Army and held numerous posts in the West during the Indian Wars. By the time of the Spanish-American War, Shafter weighed in at over 300 pounds.[12]

The Tampa Ordnance Depot was established on April 26, 1898, by Special Order No. 97, Headquarters of the Army. The commanding officer of the depot was Captain John T. Thompson. Thompson arrived in Tampa on April 28 and went straight to work to establish the depot. With only one pier and one single-track railroad into Tampa, major logistical headaches were in store for Thompson. On May 7, the Ordnance Depot in Tampa started to load the arms and ordnance stores aboard the transports scheduled to depart for Cuba. By the first of June, all ordnance stores had been loaded aboard the transports except for the reserve .30 caliber ammunition, which was loaded prior to the troops' departure. Thompson stated in his report on the operations at Tampa that the regular infantry and cavalry arrived on site well supplied with arms and equipment for their authorized peacetime strength. The task for Thompson was to arm and equip the men that were added to these regiments in order to bring them up to wartime strength. He also had to arm the volunteers that came through. When the troops left for Cuba, each man armed with a Krag had the equivalent of 500 rounds of .30-40 ammunition. Each soldier issued a Trapdoor had been supplied with 1,000 rounds of .45-70s.[13]

In Captain Thompson's September 9 report to the Ordnance Department, he stated that the follow-

The 9th Cavalry shown at Tampa, Florida, with their Krag carbines and their battle flags prior to their departure for Cuba.
U.S. Military History Institute

ing issues were made from the Tampa Depot:

Tampa Ordnance Depot[14]
Issues for the Period May-August 1898

Magazine Rifles (Krag) 4,000
Magazine Carbines (Krag) 3,000
Springfield Rifles 32,000
Revolvers cal. .38 3,000
Revolvers cal. .45 1,000
Rifle ball cartridges cal. .45 17,000,000
Rifle ball cartridges cal. .38 10,000,000

The above totals are estimates

CUBA

June-July 1898

The Army started loading troops and equipment aboard the transports for their departure to Cuba on June 11. The loading procedure, which should have taken about eight hours, actually took four days. When the 32 transport ships finally did depart, over 8,000 troops had to be left behind due to a lack of transport space. In the cavalry, each regiment left four troops behind plus all their horses except for a few of the officers' mounts. The dismounted cavalry in Cuba consisted of the 1st, 3rd, 6th, 9th, 10th and the 1st Vol. Cavalry; Troops A, C, D and F of the 2nd Cavalry obtained horses while in Cuba. The transports arrived off the Cuban coast near Santiago on June 20 with 819 officers and 16,058 enlisted men plus 2,295 horses.[15]

The first land engagement of the campaign occurred on June 24, about four miles from Siboney, at the mountain pass called Las Guasimas. Here, Major General Joseph Wheeler's (the ex-Confederate cavalry commander) Second Cavalry Brigade consisting of the 1st, 10th and the Rough Riders attacked the rear guard of General Rubin's 1,500 man force.

At 7:10 in the morning, the cavalry came upon the Spanish. The troops were halted and the command was given by Colonel Wood for the Rough Riders to load their Krag carbines. Up to this time, Wood's men had never fired their Krags. The order

While not a clear picture, it does show Captain Mullerand, Troop E, 1st U.S. Vol. Cavalry (Rough Riders) in Cuba. Note that several of the Rough Riders are shown with their Krag Model 1896 carbines.
U.S. Army Military History Institute

was then given to move forward with care. The cavalrymen advanced in good order and steadily forced the Spanish line back. Slowly, they evacuated their entire position on the ridge. Roosevelt wrote that during the fight he took a Krag carbine from a wounded man and began firing at the Spanish position. About 30 minutes after firing had ceased, the 9th Cavalry advanced about 800 yards forward of the American position to cover the entire front. During the two-hour engagement, the Spanish had two machine guns in operation. In the battle, the Spanish suffered three officers and nine men killed and twenty-four men wounded.[16] American casualties are shown below:

American Casualties[17]
Second Cavalry Brigade
Las Guasimas, Cuba
June 24, 1898

	Killed		Wounded	
	Officers	Enlisted	Officers	Enlisted
1st Cavalry	0	7	3	5
10th Cavalry	0	1	0	10
1st Vol. Cavalry	1	7	3	28
Total	**1**	**15**	**6**	**43**

SAN JUAN HILL
July 1, 1898

Shafter's plans for July 1 called for Lawton's infantry to attack the outer defensive position before Santiago at the village of El Caney. To the left of Lawton's position, additional infantry and the dismounted cavalry were to attack San Juan Ridge of San Juan and Kettle Hills. In the attack on El Caney, the Spanish, with only 600 men armed with Mauser Rifles, held out for nine hours. They retreated only after running out of ammunition. Lawton's casualties in this action consisted of 81 killed and 360 wounded.[18]

While Lawton's attack was in progress, the cavalry and infantry attacked San Juan Ridge. Because of the extreme heat, poor communications and lack of reconnaissance of the position in their front, there was mass confusion in the American ranks. As soon as the Americans came out of the jungle before the Spanish positions on San Juan Ridge, they came under fire. At this point, the Americans suffered many casualties. In this engagement, each one of the American troops was armed with 150 rounds of Krag ammunition. With the ability to fire 42 aimed shots in a two-minute

The ex-Confederate cavalryman Joseph Wheeler led Shafter's cavalry during the Cuba Campaign of June/July 1898. He is shown here as a major general of volunteers during the Spanish-American War. In 1900, he retired as a brigadier general in the regular army.
National Archives collection

period, it did not take long before the Krags became too hot to fire. To cool them off, many soldiers poured water over them from their canteens.

After being under fire for nearly two hours, Roosevelt mounted his horse and led the Rough Riders and the 10th Cavalry in the famous charge up Kettle Hill. To help cover the cavalry's advance, the four Gatling gun battery of Lieutenant Parker gave covering fire. Part way up the hill, the Americans were stopped by strands of barbed wire that had been placed there before the war for the grazing of cattle. After laying before the wire for a few minutes and shooting at the Spanish on the crest of the hill, Roosevelt climbed over the wire and called for his men to follow. Soon the whole American line was again on the attack. Captain John Bigelow, Jr. of the 10th Cavalry wrote that they did not stop until the hill was taken, except for individual soldiers who halted to fire over the heads of the men in front of them. Unfortunately, some of these shots hit the men in front. By the time the Americans reached the

The cavalry in the process of departing for Cuba in 1898.
National Archives Collection

Lieutenant Parker's Gatling Gun Battery gave supporting fire for the dismounted cavalry attack on San Juan Ridge on July 1, 1898.
U.S. Military History Institute

top of the hill, the Spanish were already retreating to their inner trenches before Santiago. With Kettle Hill safely in American hands, San Juan Hill was quickly taken. The first unit up this hill was the 10th Cavalry. In the San Juan Ridge battle, the Americans lost 142 killed and 1,014 wounded, nearly a third of which were from the cavalry. Casualties amongst the 3,010 cavalrymen in the fight are shown in the schedule below:

U.S. Cavalry Casualties[19]
San Juan Ridge
July 1, 1898

	Killed		Wounded	
	Officers	Enlisted	Officers	Enlisted
1st Cavalry	1	12	0	47
2nd Cavalry	0	0	0	1
3rd Cavalry	0	3	6	46
6th Cavalry	0	2	4	53
9th Cavalry	1	2	2	16
10th Cavalry	2	5	9	65
1st Vol. Cavalry	1	14	4	69
Brigade Hdqts.	0	0	3	0
Total	**5**	**38**	**28**	**297**

The San Juan Ridge engagement was the last battle of the war in Cuba. By the end of July, the Spanish had surrendered the island to the Americans. The war would be taken to Puerto Rico and the Philippines, but cavalry did not play a major role in either of these theaters. On August 12, 1898, Spain and the United States entered into an armistice.

The following two charts reflect the quantities of small arms and other equipment issued to the troops during the period between April 1 and August 31, 1898:

Ordnance Department[20]
Troop Issues
Spanish-American War

	Small Arms	Ammunition
.30 cal. Krag Rifles	53,571	{17,428,568
.30 cal. Krag Carbines	11,715	
.45 cal. Springfield Rifles	84,391	39,414,168
.45 cal. Springfield Carbines	3,276	2,977,118
.38 cal. Revolvers	9,515	1,468,181
.45 cal. Revolvers	13,363	569,537
Sabres	8,045	—

This second schedule reflects cavalry equipment issued to the cavalry during the Spanish-American War:[21]

Troop B, 5th Cavalry taken stateside with their Model 1896 Krag carbines. *U.S. Military History Institute*

Carbine slings..........................4,620
Carbine sling swivels...................3,279
Pistol cartridge boxes, cal. .38............9,520
Pistol holsters, cal. .38 & .45............33,304
Saber belts............................14,399
Saber attachments......................4,836
Saber knots...........................13,491
Canteen straps — short 20,550
Carbine boots..........................9,491

PHILIPPINE INSURRECTION
1899-1902

By January 1899, the American forces around Manila numbered 21,000 men, half of which were volunteers armed with Springfield Trapdoors. Included in this total of American troops were six troops of the 4th Cavalry that had been in the country since August 21 of the previous year. The 4th, armed with their Model 1896 Krag carbines, was joined in December by a troop of the Nevada Volunteer Cavalry. In 1899, the 3rd Cavalry was also deployed to the Philippines.

On January 4, 1899, the United States announced that it was taking possession of the Philippine Islands. This action caused the Filipino leader, Emilio Aquinaldo, to revolt against American rule. On the evening of February 4, shots were exchanged between sides. The U.S. was in an undeclared war. In the first two days of fighting around Manila, the Americans suffered over 250 casualties while the Filipinos lost 500. Over the next five months of pitched battles, the Americans totally defeated the insurgent forces. American casualties were heavy, with 225 killed and 1,357 wounded.[22] Seeing that different means were needed to accomplish their goal of independence, the Filipinos settled on a campaign of guerrilla operations. To deal with this new threat, the U.S. sent large reinforcements to the Philippines. By 1901, over 70,000 Americans were in the theater. The cavalry units included in this total were the 1st, 3rd, 4th, 5th, 6th, 9th and 10th.

During the early days of the conflict, the cavalry had been armed with the Model 1896 Krag carbine. In September 1899, the Springfield Armory sent to the commanding officer, Ordnance Depot, Manila, Philippines Islands, per Order for Supplies #3938, dated July 22, one thousand five hundred Model 1898 Krag carbines.[23] These new Krags would have been issued to the 3rd and 4th Cavalry — the two regiments then in country. These 1,500 Krag Model 1898 carbines were the first arms sent from the armory to

George Christopher, 4th Cavalry, in a photo taken in March 1898. The 4th did not deploy to Cuba but did see action in the Philippine Islands.
U.S. Military History Institute, MOLLUS

the Philippines in 1899. At the end of September 1899, Springfield Armory was inventorying over 8,500 Krag carbines, of which 1,639 were Model 1896, 2,695 Model 1898 and 4,200 Model 1899.[24]

Most of the fighting was over by March 1901, but sporadic fighting would last for some time to come. Needing additional troops to occupy the various outposts throughout the Philippines, Congress authorized the formation of five additional cavalry regiments on February 2, 1901.

Regiment   **Date Organized & Location**[25]
11th Cavalry.................March 11, 1901
                             Fort Meyer, Virginia
12th Cavalry...............February 8, 1901
                          Fort Sam Houston, Texas
13th Cavalry....................May 1, 1901
                          Fort Meade, South Dakota
14th Cavalry..............February 19, 1901
                          Fort Leavenworth, Kansas
15th Cavalry..............February 12, 1901
                 Presidio, San Francisco, California

The 11th, 14th and 15th Cavalry were destined for service in the Philippines within the year. To arm

Lemuel L. Reed, 4th Cavalry with his Krag carbine — photo possibly taken in the Philippines.
U.S. Military History Institute

(above) A detachment of cavalry at inspection. The photo was taken at Calamba, Philippines, about 1903.
U.S. Military History Institute

(below) Troop D, 1st Cavalry in the Philippines in 1903 with either the Model 1896 or 1898 Krag carbine.
U.S. Military History Institute

During the two-year period, 1901-1902, over 27,000 Model 1899 Krag carbines were sent to the field for issue. A larger number of these carbines were shipped to the Philippine Islands. *Smithsonian Institute collection*

the new regiments and the regiments already in the field, Springfield Armory issued 2,202 Model 1896, 401 Model 1898 and 12,753 Model 1899 Krag Carbines during the calendar year 1901.[26] The following year, an additional 14,414 Krag Model 1899 carbines were sent from Springfield for troop issue.

Model 1899 Krag Carbines[27] Issued in 1902

January	785
February	9,678
March	72
April	4
May	818
June	36
July	0
August	12
September	0
October	1,508
November	1,501
December	0
Total	**14,414**

1902 was also the last year of mass production for the Krag carbine at Springfield. One of the popular lines to come out of the conflict in the Philippines was "Civilize 'em with a Krag."

BOXER REBELLION
1900

In 1900, a secret Chinese society known to the Westerners as the Boxers set on a path of destruction by killing hundreds of foreigners and missionaries. Those individuals who escaped fled to the British embassy in Peking, which came under attack by the Boxers. Fearing for the safety of their citizens, several countries including Britain, France, Russia, Japan and the United States sent forces to relieve the siege of the British embassy.

The U.S. force was led by Major General Adna R. Chaffee. During the Apache Wars in the 1880s, Chaf-

Major General Adna R. Chaffee led the U.S. force of about 2,500 as part of the relief column that broke the Boxers seizure of the British Embassy in Peking in August 1900. *National Archives collection*

fee had been a captain in the 6th Cavalry. Chaffee's relief force consisted of a battalion of the 15th Infantry, the 2nd Battalion of the 14th Infantry, the 9th Infantry and part of the 3rd and 5th Artillery. The cavalry was made up of eight troops of the 6th that had come directly to China from San Francisco on July 1. All the U.S. forces were armed with Krags. After several engagements with the Boxers, the relief force reached Peking on August 12. The U.S. infantry played a major role in these skirmishes, using their Krag rifles to open the way to the "Forbidden City." The siege was broken. The 6th Cavalry suffered casualties in only one action; that occurred on August 19th, near Tientsin, when six troopers were wounded.[28] A year later, when the troops left China, part of the force left behind at the American Legation included four troops of the 6th Cavalry.

By 1900, the Ordnance Department had put into place the first steps to adopt a clip-loading magazine rifle for military service. On June 19, 1903, the Model 1903 Springfield rifle was officially adopted as the service arm for both the infantry and cavalry. The "03" with its clip-fed magazine, stronger action and higher velocity cartridge made the Krag obsolete. The Krag remained in field service until 1905 when it was replaced with the "03." After 70 years, the era of the carbine in the hands of the cavalry was at an end.

Footnotes - U.S. Magazine Krag, 1892-1905

[1] Ken Waters, "The U.S. Krag," *Rifle Magazine,* March/April 1978, p. 54
[2] Major James E. Hicks, *U.S. Military Firearms,* Alhambra, CA, 1962, pp. 107-108
[3] *Ibid.*
[4] Lt. Col. William S. Brophy, *Arsenal of Freedom, The Springfield Armory 1890-1948,* Mowbray Publishing, Lincoln, RI, 1991, p. 67
[5] Lt. Col. William S. Brophy, *The Krag Rifle,* Sherman Oaks, CA, 1985, pp. 16 and 49
[6] NARG 156-106
[7] Colonel W.S. Nye, *Carbine and Lance, The Story of Old Fort Sill,* Norman, OK, 1937, p. 298
[8] James A. Sawicki, *Cavalry Regiments of the U.S. Army,* Dunfries, VA, 1985, pp. 66, 203, 214 and NARG 156-970
[9] Franklin B. Mallory, "Guns of the Rough Riders Part 1, Colt Single Action Army Revolvers," *Man at Arms,* January/February 1989, pp. 11-15
[10] Franklin B. Mallory, "Guns of the Rough Riders Part II, Krag Carbines," *Man at Arms,* July/August 1989, pp. 11-16 and *Springfield Research Service Serial Numbers of the U.S. Martial Arms, Vol. 2,* Dover, DE, 1986, p. 61
[11] NARG 156-106
[12] Donald Parks, "Pecos Bill Shafter," *Wild West,* April 1992, pp. 8-16
[13] Secretary of War Report 1898, Vol. 3, pp. 245-248
[14] *Ibid.,* p. 248
[15] French E. Chadwick, *The Relations of the United States and Spain — The Spanish-American War Vol. 2,* New York, 1911, pp. 479-480
[16] *Ibid.,* pp. 53-56
[17] *Ibid.,* Appendix B
[18] *Ibid.*
[19] *Ibid.*
[20] Secretary of War Report 1898 op. cit., p. 25
[21] *Ibid.,* pp. 25-26
[22] Col. Philip H. Shockley, *The Krag-Jorgensen Rifle in the Service,* Aledo, IL, 1960, p. 38
[23] NARG 156-106
[24] *Ibid.*
[25] James A. Sawicki, op. cit., pp. 172, 174, 176, 178, 179
[26] NARG 156-1383
[27] *Ibid.*
[28] James A. Sawicki, op. cit., p. 78

APPENDIX A

Union Regimental Summary 1861-1865

All regiments shown are Union cavalry units unless otherwise stated. While Confederate units were also partially armed with Union carbines, these units were not included in these listings, but are shown in Appendix B.

BALLARD

- Alabama: 1st
- Iowa: 2nd
- Kentucky: 3rd, 6th, 8th, 11th, 12th, 13th, 16th (4th, 13th, 30th, 37th, 45th, 52nd, 54th Mounted Infantry)
- New York: 13th Heavy Artillery
- Ohio: 7th, 12th, McLaughlin Battalion

BURNSIDE

- Alabama: 1st
- Arkansas: 2nd
- Connecticut: 1st
- Illinois: 2nd, 3rd, 6th, 9th, 10th, 11th, 12th, 14th, 15th, 16th, 17th, 36th and 92nd Mounted Infantry
- Indiana: 1st, 3rd, 4th, 6th, 9th, 10th, 13th, 16th, 65th Mounted Infantry
- Iowa: 3rd, 6th, 7th, 8th
- Kansas: 16th
- Kentucky: 1st, 3rd, 4th, 5th, 6th, 7th, 11th, (39th, 40th and 45th Mounted Infantry)
- Louisiana: 1st, 2nd
- Maine: 1st, 2nd
- Maryland: 1st, 3rd
- Michigan: 1st, 3rd, 6th, 7th, 8th, 9th, 10th
- Minnesota: 2nd, Brackett's Battalion
- New Hampshire: 1st
- New Jersey: 1st, 3rd
- New York: 1st, 1st Vet., 2nd, 3rd, 4th, 6th, 8th, 9th, 10th, 11th, 12th, 14th, 15th, 16th, 18th, 21st, 22nd, 25th
- Texas: 1st, 2nd
- Ohio: 1st, 2nd, 3rd, 4th, 4th Ind. Bn., 5th, 6th, 7th, 8th, 12th, Adams Ind., McLaughlin Bn.
- Pennsylvania: 1st, 3rd, 7th, 8th, 9th, 11th, 12th, 13th, 14th, 15th, 16th, 17th, 18th, 20th, 21st, 22nd
- Rhode Island: 1st, 2nd, 3rd Cavalry plus 1st Infantry
- Tennessee: 1st, 2nd, 5th, 7th, 8th, 10th, 2nd Mounted Infantry and 1st Ind. Co.
- West Virginia: 1st, 3rd, 7th
- Wisconsin: 1st, 2nd, 3rd, 4th
- United States -
  - Regulars: 3rd, 4th
  - Colored: 1st, 4th

COSMOPOLITAN
(GWYN & CAMPBELL)

- Arkansas: 2nd, 3rd
- Illinois: 4th, 5th, 6th, 7th
- Indiana: 3rd
- Iowa: 4th, 8th
- Kansas: 2nd, 6th, 14th
- Kentucky: 11th, 12th, 14th and 40th Mounted Infantry
- Michigan: 10th
- Missouri: 6th, 8th and 3rd State Militia
- Ohio: 5th Ind. Bn., 8th
- Tennessee: 7th
- Wisconsin: 3rd
- United States -
  - Colored: 3rd, 6th

GALLAGER

- Arkansas: 2nd, 3rd
- Colorado: 2nd
- Illinois: 3rd*, 9th, 12th, 13th, 83rd Mounted Infantry
- Indiana: 3rd, 6th, 9th, 11th*
- Iowa: 1st, 4th, 7th
- Kansas: 9th, 10th, 11th, 14th, 16th
- Kentucky: 1st, 4th, 8th, 15th
- Mississippi: 1st
- Michigan: 10th
- Missouri: 4th, 12th*, 4th State Militia

Appendix A

New York: 1st
Ohio: 2nd, 2nd Ind. Bn., 3rd, 4th, 6th, 7th, 9th, 10th
Pennsylvania: 5th, 9th, 13th, 20th, 21st, 22nd
Tennessee: 1st, 2nd, 4th, 7th, 8th, 9th, 12th*, 1st and 4th Mounted Infantry
West Virginia: 5th
Wisconsin: 3rd
United States -
Colored: 3rd

* *Units armed with the rimfire .56—.50 Gallager carbine*

GIBBS

Missouri: 10th
New York: 13th, 16th

HALL

Arkansas: 4th
Dakota Territory: 1st
Illinois: 3rd, 4th, 9th, 10th, 11th
Indiana: 1st
Iowa: 3rd, 5th and Sioux City Iowa Cavalry
Kansas: 2nd, 5th, 6th, 11th, 15th
Missouri: 2nd, 3rd, 4th, 6th, 7th, 10th, (2nd, 3rd, 6th, 12th State Militias)
Ohio: 11th, Bracken's Bn.
New York: 8th
Wisconsin: 2nd

JOSLYN

California:	2nd	M1864
Indiana:	4th, 8th	"
Kentucky:	2nd	"
Missouri:	4th	"
Nebraska:	1st	"
New York:	19th	M1862
Ohio:	2nd, 3rd, 4th, 6th, 11th (all percussion except 11th M1864)	
Pennsylvania:	9th	M1864
Tennessee:	13th	M1864
West Virginia:	2nd, 3rd	"
Wisconsin:	1st	"
United States - Colored:	3rd, 5th	"

LINDNER

Michigan: 1st Austrian Lindners

West Virginia: 7th Cavalry and 8th Mounted Infantry (Amoskeag delivery of 1/63)

MAYNARD

California: 2nd - Wartime Model
Indiana: 6th, 11th - Wartime Model
Kansas: 5th, 6th, 7th, 9th Pre-war delivery 12/57
Pennsylvania: 9th - Pre-war delivery 12/57
Tennessee: 10th, 12th - Wartime Model
Wisconsin: 1st - Pre-war delivery 12/57
United States -
Regulars: 4th - Pre-war delivery 12/57
1st, 3rd, 4th - Wartime Model

MERRILL

Arkansas: 2nd
Colorado: 2nd
Delaware: 1st
Illinois: 1st, 11th, 16th, 83rd Mounted Infantry
Indiana: 2nd, 6th, 7th, 12th
Kansas: 2nd, 6th, 7th, 14th
Kentucky: 2nd, 5th, 4th 26th Mounted Infantry
Maryland: Purnell Legion
Michigan: 3rd
Missouri: 9th, 11th
Nebraska: 1st
New Jersey: 1st
New York: 1st, 1st Vet., 4th, 5th, 18th
Ohio: 3rd, 10th, 11th
Pennsylvania: 5th, 11th, 17th, 18th
Tennessee: 1st, 2nd, 3rd, 5th, 10th, 11th, 12th, 1st and 2nd Mounted Infantry
Wisconsin: 1st, 3rd
United States -
Colored: 2nd

MUSKETOON

Arizona: 1st
Colorado: 1st
Kansas: 7th
Kentucky: 1st
Louisiana: 2nd
Michigan: 3rd, 8th
Missouri: 1st, 3rd, 12th, 13th State Militias
Nebraska: 1st

Ohio: 11th
Texas: 1st
Wisconsin: 3rd

M1855 PISTOL CARBINE

Illinois: 9th
Indiana: 1st
Iowa: 4th
Kansas: 2nd, 6th, 9th
New York: 11th
West Virginia: 1st, 2nd

SHARPS

Arkansas: 2nd, 3rd, 4th
California: 1st, 2nd, 1st Bn.
Colorado: 1st, 3rd
Dakota Territory: 1st
Illinois: 1st, 2nd, 3rd, 4th, 5th, 6th, 7th, 8th, 9th, 10th, 11th, 14th, 15th, 16th, 17th
Indiana: 1st, 3rd, 4th, 5th, 6th, 8th
Iowa: 1st, 2nd, 3rd, 4th, 5th, 6th, 9th
Kansas: 2nd, 5th, 6th, 7th, 9th, 11th, 14th, 15th
Kentucky: 1st, 2nd, 3rd, 4th, 6th, 7th, 11th, 12th
Louisiana: 1st
Maryland: 1st, 2nd, 3rd, Upper Potomac Force
Massachusetts: 1st, 2nd, 3rd, 4th, 40th Mounted Infantry
Michigan: 1st, 2nd, 3rd, 4th, 5th, 8th
Mississippi: 1st
Maine: 1st
Minnesota: 1st, 2nd
Missouri: 1st, 2nd, 3rd, 4th, 6th, 7th, 10th, 11th, 15th, 16th, 32nd Mounted Infantry, 1st, 3rd, 7th and 8th State Militias
New Hampshire: 1st, 8th Infantry
New Mexico: 1st
Nevada: 1st
Nebraska: 1st
New York: 1st, 1st Lincoln, 1st Vet., 1st Mtd. Rifles, 2nd, 2nd Vet. 2nd Mtd. Rifles, 3rd, 4th, 5th, 6th, 7th, 8th, 9th, 10th, 13th, 14th, 15th, 16th, 19th, 20th, 22nd, 24th
New Jersey: 1st, 2nd, 3rd
Ohio: 1st, 2nd, 3rd, 4th, 5th, 6th, 10th, 13th, Adams Ind. Bn.
Oregon: 1st
Pennsylvania: 1st, 2nd, 3rd, 4th, 5th, 6th, 8th, 9th, 11th, 12th, 13th, 15th, 16th, 17th, 19th, 21st, 22nd
Rhode Island: 1st
Tennessee: 1st, 2nd, 3rd, 6th, 10th, 11th, 12th, 3rd Mounted Infantry
Texas: 1st, 2nd
Vermont: 1st
West Virginia: 1st, 2nd, 3rd
Wisconsin: 1st, 2nd, 4th
United States -
Regulars: 1st, 2nd, 3rd, 4th, 5th, 6th
Colored: 3rd, 4th, 5th

SHARPS & HANKINS

New York: 2nd Vets., 3rd, 9th, 10th, 11th

SMITH

Alabama: 1st
Arkansas: 2nd
Connecticut: 1st
Illinois: 2nd, 7th, 11th
Indiana: 2nd, 3rd, 4th, 5th, 65th Mounted Infantry
Kansas: 2nd, 9th, 10th, 11th, 14th, 16th
Kentucky: 1st, 2nd, 3rd, 4th, 5th, 6th, 7th, 8th, 9th, 11th, 30th Mounted Infantry
Maine: 1st
Maryland: 1st, 2nd
Massachusetts: 1st
Michigan: 10th
Minnesota: 1st, 2nd, Ind. Co., 8th Mounted Infantry
Missouri: 5th Militia, 13th Vol. 7th Mounted Infantry
New Jersey: 1st
New York: 1st, 1st Vet., 2nd, 9th, 10th, 18th, 21st
Ohio: 6th, 9th, 11th, McLaughlin Bn.
Pennsylvania: 7th, 17th
Tennessee: 1st, 5th
West Virginia: 1st, 2nd, 3rd, 4th, 5th
Wisconsin: 3rd
United States -
Regulars: 1st, 2nd, 3rd, 4th, 5th
Colored: 3rd, 4th

SPENCER

Florida: 1st

Connecticut: 1st
Delaware: 1st
Illinois: 2nd, 3rd, 6th, 9th, 12th, 17th
Indiana: 2nd, 4th, 7th, 8th, 9th
Iowa: 2nd, 3rd, 4th, 5th, 8th
Kansas: 6th, 7th, 14th
Kentucky: 2nd, 3rd, 4th, 6th, 7th, 11th, 4th Mounted Infantry
Maine: 1st, 2nd
Massachusetts: 1st, 2nd, 4th
Michigan: 1st, 2nd, 3rd, 4th, 5th, 6th, 7th, 9th, 10th, 11th
Missouri: 8th, 10th, 9th State Militia
New Hampshire: 3th and 7th Infantry
New Jersey: 1st, 2nd, 3rd
New York: 1st, 1st Vet., 1st Mounted Rifles, 2nd, 5th, 10th, 13th, 18th, 19th, 21st, 22nd, 25th
Ohio: 1st, 2nd, 3rd, 4th, 5th, 7th, 8th, 9th, 10th, 11th, 12th
Pennsylvania: 4th, 5th, 7th, 9th, 11th, 13th, 14th, 15th, 16th, 17th, 18th
Rhode Island: 1st
Tennessee: 2nd, 12th
Vermont: 1st
West Virginia: 1st, 2nd, 3rd, 5th, 6th, 7th
Wisconsin: 1st, 2nd
United States -
Regulars: 1st, 2nd, 4th, 6th

STARR

Arkansas: 1st, 3rd
Colorado: 1st, 2nd, 3rd
Illinois: 13th
Iowa: 9th
Kansas: 5th, 7th, 11th, 16th
Kentucky: 1st, 30th Mounted Infantry
Michigan: 3rd, 10th, 11th
Missouri: 2nd, 7th, 11th, 12th, 13th, 14th (6th, 7th, 12th State Militias)
New Jersey: 1st, 3rd
New York: 1st Vet., 2nd, 4th, 9th, 12th, 15th, 20th, 23rd, 24th
Nevada: 1st
Ohio: 2nd, 9th, 11th
Pennsylvania: 9th, 13th, 16th, 17th, 19th, 22nd (12th & 14th with rimfire Starrs
Tennessee: 19th
Vermont: 1st
West Virginia: 1st, 6th
United States -
Regulars: 3rd, 2nd in 1866 with rimfire Starrs

WARNER

Colorado: 1st
Massachusetts: 3rd
Wisconsin: 1st

WESSON

Kentucky: 7th, 8th, 9th
Missouri
State Militia: 4th, 6th, 8th
Ohio: 11th Infantry

APPENDIX B

Confederate Regimental Summary

The Confederate cavalry was armed with a large variety of small arms ranging from shotguns to breechloading carbines and rifles. Most of the breechloaders were captured from the Union forces with only the First Model Maynard being totally state-purchased prior to the outbreak of hostilities in April 1861. Other arms, such as small quantities of Merrill and Sharps carbines, were also purchased prior to the war. The other carbines reflected in this list were captured arms. Most Southern cavalry regiments were heavily armed with muzzle loading muskets and rifles and were issued only a small portion of breechloaders.

BALLARD
Texas: 3rd, 9th
CSA-Regulars: 6th

BURNSIDE
Louisiana: 3rd
Georgia: 8th
Mississippi: 18th, 19th
North Carolina: 1st, 2nd, 3rd, 5th
Texas: 2nd, 8th, 9th
Tennessee: 9th
Virginia: 4th, 6th, 7th, 9th, 10th, 11th, 12th, 35th Bn., 45th
CSA-Regulars: 1st, 3rd, 8th

COSMOPOLITAN
Mississippi: 19th
Texas: 3rd, 6th, 9th, 27th
Tennessee: 9th

COLT CARBINE
Mississippi: 18th
Tennessee: 9th

GALLAGER
Texas: 6th
Tennessee: 9th
Virginia: 11th

HALL
Florida: 2nd
North Carolina: 1st, 2nd
Mississippi: 19th
Texas: 11th
Tennessee: 9th
Virginia: 1st, 6th, 11th, 12th

MAYNARD
Georgia: 5th, 7th, 9th, 20th, Cobbs Legion
Florida: 2nd, (1st Spec. Bn. Inf. & 6th Sp. Bn. Inf.)
Louisiana: 1st, 3rd, (11th Inf.)
Mississippi: 1st, 4th, 18th, 19th, (9th, 14th, 15th Inf.)
Tennessee: 3rd, 9th
Virginia: 35th Bn.
North Carolina: 18th Inf.

MERRILL
Georgia: 8th
North Carolina: 1st, 2nd, 3rd, 5th, 6th
Texas: 1st, 27th
Tennessee: 7th, 17th
Virginia: 1st, 7th, 10th, 11th, 12th, 13th, 14th, 35th Bn.

SHARPS
CSA-Regulars: 1st, 3rd, 7th, 8th
Georgia: 4th, 5th, 7th, 8th, 11th, 20th, 59th, 2nd State Troops Inf., Jeff Davis Legion, Phillips Legion, Cobbs Legion
Louisiana: 3rd
Mississippi: 2nd, 11th, 18th, 19th, 26th, 42nd
North Carolina: 1st, 2nd, 3rd, 4th, 5th, 16th
South Carolina: 2nd
Tennessee: 3rd, 9th
Texas: 3rd, 6th, 8th, 9th, 27th
Virginia: 1st, 5th, 6th, 7th, 8th, 9th, 10th, 12th, 13th, 24th, 30th, 35th Bn.

SMITH
Louisiana: 3rd
North Carolina: 1st, 2nd, 3rd, 5th
South Carolina: 7th
Texas: 6th, 9th
Virginia: 7th, 11th, 12th, 13th, 35 Bn.

SPENCER
North Carolina: 1st, 2nd, 3rd, 5th
Texas: 3rd, 6th, 8th, 9th, 27th
Virginia: 3rd, 9th, 10th, 13th, 43rd Bn.
CSA-Regulars: 6th

BIBLIOGRAPHY

PRINTED BOOKS

Annual Report, The Quartermaster General to the Governor of the State of Kentucky for 1863-64, Frankfort, 1865

Battles and Leaders of the Civil War, A.S. Barnes & Co., Inc., New York, 1956

Brophy, William S. Lt. Col., *Arsenal of Freedom, The Springfield Armory 1890-1948*, Andrew Mowbray Inc., Lincoln, RI, 1991; *The Krag Rifle*, Beinfield Publishing Inc., Sherman Oaks, CA, 1985

Coates, E.J. & Thomas, D.S., *An Introduction to the Civil War Small Arms*, Thomas Publications, Gettysburg, PA, 1990

Chadwick, French E., *The Relations of the United States and Spain – The Spainish-American War Vol. 2*, Charles Scribner's Sons, New York, 1911

Davis, W.C., *The Fighting Men of the Civil War*, Gallery Books, New York, 1989

Dyer, F.H., *A Compendium of the War of the Rebellion*, Morningside Bookshop, Dayton, OH, 1978

Edwards, W.B., *Civil War Guns*, Stackpole Co., Harrisburg, PA, 1962

Executive Document, State of Ohio, 1862

Executive Document, No. 99, 40th Congress, 2nd Session, 1868

Frasca, Albert J. and Hall, Robert H., *The .45-70 Springfield*, Springfield Publishing Co., Northridge, CA, 1980

Fuller, Claud E., *The Breech-Loader in the Service 1816-1917*, N. Flayderman & Co., New Milford, CT, 1965

Garavalgia, L.A. and Worman, C.G., *Firearms of the American West 1803-1865*, University of New Mexico Press, Albuquerque, NM, 1984; *Firearms of the American West 1866-1894*, University of New Mexico Press, Albuquerque, 1985

Genco, J.G., *Arming Michigan Regiments 1862-1864*, 1982

Hard, M.D., *History of the 8th Cavalry Regiment Illinois Volunteers*, Morningside Bookshop, Dayton, OH, 1984

Hicks, Major James E., *U.S. Military Firearms*, Borden Publishing Co., Alhambra, CA, 1962

Hull, E.A., *The Burnside Breechloading Carbine*, Andrew Mowbray Inc., Lincoln, RI, 1986

Mallory, Frank B., *Serial Numbers of U.S. Martial Arms Vol. 2*, Springfield Research Service, Dover Litho Printing Co., Dover, DE, 1986

Marcot, R.M., *Spencer Repeating Firearms*, Northwood Heritage Press, Irvine, CA, 1983

McAulay, J.D., *Carbines of the Civil War 1861-1865*, Pioneer Press, Union City, TN, 1981; *Civil War Breechloading Rifles*, Andrew Mowbray Inc., Lincoln, 1987; *Civil War Carbines* Vol. II, Andrew Mowbray Inc., Lincoln, 1991

Minnigh, L.W., *Gettysburg: What They Did Here*, Bookmark Publishers, Gettysburg, PA, 1954

Nye, Col. W.S., *Carbine and Lance – The Story of Old Fort Sill*, University of Oklahoma Press, Norman, OK, 1937

Official Records of the Rebellion, 128 Vols., GPO; 1891

Reid, W., *Ohio in the War*, Cincinnati, 1868

Reilly, R.M., *United States Military Small Arms 1816-1865*, Eagle Press Inc., Baton Rouge, LA, 1970

Sarf, Wayne M., *Great Campaigns – The Little Bighorn Campaign March-September 1876*, Combined Books Inc., Conshohocken, PA, 1993

Sawicki, J.A., *Cavalry Regiments of the U.S. Army*, Wyvern Publications, Dumfries, VA, 1985

Scott, Douglas D. and Fox, Richard A. Jr., *Archaeological Insight into the Custer Battle*, Norman, OK, 1987

Secretary of War Reports Volume 3, 1879, 1880, 1881, 1882, 1886 and 1898

Sellers, F., *Sharps Firearms*, Beinfield Publishing Inc., North Hollywood, CA, 1978

Shockley, Col. Philip H., *The Krag Jorgensen Rifle in the Service*, World-Wide Gun Report Inc., Aledo, IL, 1960

Starr, S.Z., *The Union Cavalry in the Civil War, Vols. I, II, III*, Louisiana State University Press, Baton Rouge, LA, 1979, 1981, 1985

Todd, F.P., *American Military Equipage, 1851-1872 Vol. II State Forces*, Chatham Squire Press, Inc., New York, 1981, *American Military Equipage, 1851-1872 Vol. I*, Charles Scribner's, New York, 1980

Thomas, D.S., *Ready Aim Fire*, Thomas Publications, Biglerville, PA, 1981

Utley, R.M., *Frontiersmen in Blue*, University of Nebraska Press, Lincoln, NE, 1967; *Frontier Regulars*, University of Nebraska Press, Lincoln, NE, 1973

Warner E.J., *Generals in Blue*, Louisiana State University Press, Baton Rouge, LA, 1964

West, E.C., *History and Reminiscences of the Second*

Wisconsin Cavalry Regiment, Grand Army Press, Rochester, MI, 1982

Williams, T. Harry, *The History of American Wars from 1745-1918*, Alfred A. Knoff Inc., New York, 1981

MAGAZINE ARTICLES

Albaugh, W.A. III, "Union Armament in the Civil War," *North/South Trader*, March/April 1975

Austerman, W.R., "Seven Round Rimfire – The Spencer on the Frontier," *Man at Arms*, March/April 1985; "The Spencer in Confederate Service," *Arms Gazette*, September 1980; "Maza Wakan for the Sioux," *Man at Arms*, January/February 1988

Houze, H.G., "Schuyler, Hartley & Graham Military Sales to France During the Franco-Prussian War 1870-1871," *The Gun Report*, June 1993

Hull, E.A., "The Burnside Breechloading Carbine," Lincoln; 1986, "The Starr-Woolcott Carbines," *Man at Arms*, May/June, 1984

Hutchins, James S., "Captain Michaelis Reports on Army Weapons and Equipment on the Northern Plains 1876-1879," *Man at Arms*, January/February 1988

Jackson, E.J., "A Gibbs Carbine and Pvt. Milton F. Murphy," *North/South Trader*, September/October 1986

Jordan, Robert P., "Ghosts on the Little Bighorn," *National Geographic*, December 1986

Killblane, Richard E., "Arizona Tiger Hunt," *Wild West*, December 1993

Lustyik, A.F., "Triplett & Scott," *The Gun Report*, June/July 1979; "The Joslyn Carbine," *The Gun Report*, September/October 1962; "The Warner Carbine," *The Gun Report*, May 1960

Marcot, R.M., "Springfield Armory Conversions and Repairs to Spencer Repeating Carbines," *The Gun Report*, July 1980; "Remington Split-Breech Carbines," *The Gun Report*, February 1991

McAulay, John D. "Arming the Union Cavalry at Gettysburg," *The Gun Report*, September 1980, "Carbines for General Buford's Cavalry at Gettysburg," *The Gun Report*, January 1979

Mallory, Frank B., *Guns of the Rough Riders Part 1*, Man at Arms, January/February 1989, *Guns of the Rough Riders Part 2*, Man at Arms, July/August 1989.

Parks, Donald, *Pecos Bill Shafter*, Wild West, April 1992

Smith, R.B., "Biggest Indian Fight," *The Wild West*, April 1992

Waters, Ken, "The U.S. Krag," *Rifle Magazine*, March/April 1978

NATIONAL ARCHIVES

<u>Records of the Chief of Ordnance, Record Group 156</u>

Section 21. Letters Received 1812-94

Section 100. Quarterly Statements of Ordnance on Hand at Forts 1862-1864

Section 101. Quarterly Statements of Ordnance on Hand Arsenal & Depot 1862-63

Section 103. Weekly Statements of Ordnance on Hand Arsenal & Depot 1864

Section 105. Monthly Statements of Ordnance Stores on Hand at Ordnance Depots Jan.-June 1890

Section 106. Monthly Statements of Ordnance on Hand at Arsenals July 1897-Feb. 1902

Section 108. Inventories of Ordnance as of October 24, 1862

Section 110. Quarterly Statements of Ordnance in Cavalry 1862-64

Section 112. Quarterly Statements of Ordnance in Hands of Troops 1872-76

Section 118. Ordnance Issued to the Militia 1816-1904

Section 124. Sales of Ordnance at Depots & Arsenals 1864-1907

Section 125. Sales of Ordnance at Depots & Arsenals 1865-94

Section 215. Abstracts of Army Officers Reports on Small Arms 1863-64

Section 970. Notes, Memoranda, and Letters Received Relating to the Spanish American War 1898

Section 1354. Letters and Endorsements Sent to the Chief of Ordnance Jan. 1873-April 1895

Section 1362. Letters Received (Miscellaneous) 1880-97

Section 1383. Weekly Reports of Arms Manufactured, Repaired, Issued, and Ready for Issue 1883-86; 1901-04

<u>Inventory of the Records of the Accounting Offices of the Dept. of the Treasury, Record Group 217</u>

Section 759. State Claims for Reimbursement of Civil War Expense Ca. 1861-1900

<u>Records of the Department of the Army, Record Group 108</u>

Section 75. Field Reports on Efficiency of Carbines

Section 77. Ordnance Stores in the Cavalry

<u>Records of the Confederate States of America, Record Group 109</u>

Entry M935. 18 Rolls-Inspection Reports 1864

INDEX

A

Adams Independent Cavalry, 63, 64
Adobe Walls, 41, 50
Alabama 1st Cavalry (U.S.), 27, 48, 52
Allegheny Arsenal, 75, 76, 81, 86-88, 92
Allen pistols, 11
Ames Model 1860 naval cutlasses, 67
Amoskeag Mfg. Co., 23
Angel Island (California), 88
Apache, Fort (Arizona), 110
Apache Indians, 50, 96, 106
Apache Pass Station, 12
Apache Wars, 107-111, 130
Appomattox Campaign, 58
Appomattox Courthouse, 60
Aquinaldo, Emilio, 127
Argentina, 76
Arizona, State of, 12
Arizona Territory, 88
Arkansas Cavalry:
    1st, 52
    2nd, 35, 44, 52, 63
    3rd, 52
    4th, 52
Ashville rifles, 51
Atlanta Campaign, 48, 50
Augusta Arsenal, 97
Austrian rifles, 51

B

Baldwin, Austin & Co., 78
Ball carbines, 69-71, 78, 80, 86
Ballard carbines, 13, 14, 17, 18, 34, 35, 39, 41-43, 48, 51, 57, 58, 67, 68, 75, 77, 79, 80, 86, 92, 93
Ballard cartridges, 81, 94
Ballard musketoons, 68
Ballard rifles, 61, 79
Baltimore & Ohio Railroad, 60
Bannerman, Francis, 80, 81
Bascom, Fort (New Mexico), 50
Baton Rouge Arsenal, 86, 87, 92
Bayant, Capt. Asa, 41
Bayard, Fort (New Mexico), 107
Beal Remington pistols, 11
Beecher's Island, 90
Belgian rifles, 51
Benicia Arsenal, 69, 81, 86-88, 92, 97, 104, 106, 110
Benteen, Capt. Frederick, 102, 104
Berdan's sharpshooters, 45
Bigelow, Capt. John Jr., 124
Black Hills Expedition, 98
Blakeslee accouterments, 78
Blakeslee cartridge boxes, 61, 65, 66
Blakeslee, Erastus, 65
Boker, H. & Co., 76, 79, 80
Boxer Rebellion, 130
Brackett Battalion, 53, 63, 64
Brandy Station, 29, 30, 55
Bristol Firearms Co., 10, 12
Brown, Capt. W.C., 120
Buffington rear sights, 108, 110
Buford, Brig. Gen. John, 29-31
Bull Run, 11
Burnside carbines, 10-12, 15, 17, 24, 30-35, 38, 39, 41-43, 46-49, 51, 57-64, 66, 67, 75, 79-81, 86, 93
Burnside cartridges, 81, 94
Burnside (gunboat steamer), 68

C

California Cavalry:
    1st, 12, 19, 35, 50, 52
        1st Battalion Native at San Francisco, 35
    2nd, 12, 19, 35, 52, 57, 63, 68, 69
    2nd Volunteer, 12
California National Guard, 92
Canada, 70
Canby, Brig. Gen. Edward R.S., 96
Capehart, Brig. Henry, 49, 59
Captain Jack, 96
Carlton's California Column, 12
Carr, Capt. Eugene A., 11
Carson, Col. Christopher "Kit", 41, 50, 51
Cavalry (See individual states or U.S. Cavalry)
Cedar Creek, 50
Chaffee, Maj. Gen. Adna R., 130
Chaffee-Reese rifles, 107
Chapman, Col. George, 47, 49
Charleston Arsenal, 86, 87
Cheyenne Indians, 90, 104
Cheyenne Ordnance Depot, 72, 110
Chickamauga Park (Georgia), 34, 120
Clapp, Everett, 70
Cold Harbor, 46
Colorado Cavalry:
    1st, 19, 35, 52, 72
    2nd, 19, 52
    3rd, 52, 67
Colt carbines, 51, 55, 57, 58, 61
Colt cartridges, 94
Colt Dragoon pistols, 11
Colt revolvers, 13, 14, 17, 30, 45, 55, 88, 105
Colt rifles, 13, 17, 29, 34, 61
Colt Single Action revolvers, 118, 120
Columbus Arsenal, 86, 87, 92
Comanche Indians, 50, 90, 102
Connecticut 1st Cavalry, 19, 35, 47, 49, 60, 64, 65
Cosmopolitan carbines, 13, 14, 17, 29, 31, 35, 39, 44, 45, 48, 57, 58, 67, 86, 93
Cosmopolitan cartridges, 94
Crook, Lt. Col. George, 58, 59, 89, 97
Cuba, 120, 121, 123, 126, 127
    (Las Guasimas) 123, 124
Custer, Capt. Thomas, 105
Custer's 7th Cavalry, 88, 90, 98, 102, 103, 105, 108
Custer, Gen. George A., 31, 32, 45, 46, 49, 50, 58-60, 102-106

D

Dakota 1st Cavalry, 35, 52, 64
Dana's Independent Co. Cavalry, 36
Davidson, Col. John, 102
Davies, 1st Brig. Gen. Henry, Jr., 47, 58, 59
Davis, Fort (Texas), 111
Dead Man's Fork, 69
Delaware 1st Cavalry, 35, 52, 60
Denvor and Pack, 24
DeSilver, I.F. & Son, 76
Detroit Arsenal, 87
Devin, Col. Thomas, 32, 46, 49, 58, 59
Dragoons (1st), 11, 12
Dragoons (2nd), 11, 12
Droop Mountain, 23
Dyer, Gen. A.B., 78

E

Ellis, Fort (Montana), 102
Enfield rifles, 23, 34, 44, 51, 68, 105
Ewell, Lt. Gen. Richard, 60
Experimental carbines, Model 1870, 97, 98, 101, 102
Experimental carbines, Model 1886, 108, 110

F

Farger, Capt. Dan, 29
Farnsworth, Elon, 31, 32
Fetterman, Capt. William J., 88
Fetterman Massacre, 88
First Army Corps of Infantry, 31
Fisher's Hill, 50
Fitzhugh, Charles, 59
Five Forks, 60
Florence, Col. Thomas, 14
Florida 1st Cavalry, 64
Fogarty Rifle Co., 92
Folson, Charles, 76
Forrest, Nathan B., 12, 51, 62
Forsyth, Col. James W., 112
Foster (gunboat steamer), 68
Franco-Prussian War, 77
Frankford Arsenal, 68, 86-88, 92
French pistols, 51
Frohock, Capt. William, 89

G

Gallager carbines, 11, 14, 15, 17, 26, 30, 32-35, 39, 43, 44, 46, 47, 51, 57-59, 67, 68, 70, 75, 77, 78, 80, 81, 86, 93
Gallager cartridges, 81, 94
Gansevoort, Lt. Col., 26
Gatling gun, 124, 125
Geronimo, 107-109, 120
Gettysburg, Battle of, 28-31, 33
Ghost Dance, 110, 112
Gibbs carbines, 25, 35, 39, 75, 80, 86, 93
Gibbs, Col. Alfred, 24, 33, 59
Godfrey, C.J., 81
Grant, Gen. Ulysses S., 29, 45, 46, 60, 70, 88
Grant, L.A., 119
Gravelly Springs, 61
Greene carbines, 12, 15, 35, 39
Gregg, Brig. Gen. David McMurtrie, 30, 31, 47
Gregg, J. Irvin., 32, 47, 59
Grierson, Col. Benjamin H., 29
Grierson's 10th Cavalry, 88, 106
Grierson's Raid, 29, 31
Gwyn & Campbell carbines, 13, 44, 45, 75, 81

H

Hall carbines, 11, 12, 15, 17, 24, 25, 35, 39, 51, 57, 58, 75, 80, 86, 93
Hall cartridges, 94
Hamilton, Col. William, 66
Hampton, Wade, 30, 45, 46
Hancock Station, 58
Harpers Ferry (West Virginia), 75, 88
Hartley, Marcus Co., 71, 80, 81
Hatch, Edward, 61
Haw's Shop, 46
Henry rifles, 14, 59, 88, 105
Hill, A.P., 31
Hilton Head Arsenal, 86, 87
Holster pistols, 51
Horseshoe Canyon, 107
Hotchkiss carbines, 105-107, 111
Huachuca, Fort (Arizona), 109
Huff, John A., 45

I

Illinois Cavalry:
  1st, 26
  2nd, 19, 35, 52, 64
  3rd, 11, 19, 24, 30, 35, 52, 61, 63, 64, 67, 68
  4th, 19, 35, 41, 52, 63
  5th, 19, 35, 41, 45, 52, 63
  6th, 19, 29, 35, 52, 61, 64, 66, 67
  7th, 19, 29, 35, 52, 61, 63
  8th, 19, 30-32, 35, 47
  9th, 19, 35, 52, 61, 63, 64, 66, 67
  10th, 19, 35, 52, 63, 64
  11th, 19, 35, 52, 64
  12th, 19, 32, 35, 52, 64
  13th, 35, 44, 52
  14th, 34, 35, 48, 52, 64
  15th, 35, 52
  16th, 35, 48, 52, 64
  17th, 63, 64
Illinois Mounted Infantry:
  92nd, 48, 66
  98th, 48, 61, 62
  123rd, 48, 61, 62
  Steward Battalion, 21
Independent Co. Cavalry (1st), 37
Indiana Cavalry:
  1st, 19, 24, 35, 52, 63
  2nd, 35, 48, 52, 61, 64
  3rd, 19, 32, 34, 35, 41, 47, 48, 52, 63
  4th, 35, 48, 52, 61, 63, 64
  5th, 34, 35, 48, 52, 63
  6th, 35, 48, 52, 68
  7th, 26, 52, 65
  8th, 48, 52, 63, 65, 66
  9th, 52, 65
  10th, 42, 52, 64
  11th, 52, 61, 67, 68
  12th, 52
  13th, 64
Indiana Mounted Infantry:
  16th, 64
  17th, 48, 61, 62
  65th, 35
  72nd, 48, 61
Indiana, State of, 34
Indianapolis Arsenal, 87, 92, 94
Iowa Cavalry:
  1st, 19, 41, 52, 63
  2nd, 19, 29, 35, 52, 61, 65, 67
  3rd, 19, 35, 52, 61, 62, 65
  4th, 11, 35, 40, 52, 61, 62, 65
  5th, 19, 35, 48, 52, 61, 65
  6th, 52, 64
  7th, 35, 52, 64
  8th, 44, 48, 52, 61, 65
  9th, 42, 52, 63

J

Jenks carbines, 14
Jorgensen, Erik, 119
Joslyn carbines, 11, 14, 24, 25, 33, 35, 39, 40, 46, 48, 57, 58, 66, 72, 75, 77-81, 86, 92, 93
Joslyn cartridges, 24, 94
Joslyn revolvers, 14

K

Kansas Cavalry:
  2nd, 19, 35, 52
  5th, 19, 35, 52, 63
  6th, 19, 35, 52, 67
  7th, 19, 35, 41, 52, 65

Index

9th, 35, 43, 52, 63, 67
10th, 52
11th, 35, 52, 63
14th, 35, 52, 67
15th, 52, 63
16th, 52, 64
19th Volunteer, 90
Kansas Infantry:
    2nd, 12
Kearny, Fort (Nebraska), 88
Keneba Arsenal, 81
Kenneche Arsenal, 86, 87, 92
Kentucky Cavalry:
    1st, 34, 35, 48, 52
    2nd, 19, 26, 34, 35, 48, 52, 63, 65, 66
    3rd, 19, 35, 48, 52, 62, 64-66
    4th, 34, 35, 48, 52, 61, 65
    4th Volunteer, 26
    5th, 19, 34, 35, 48, 52
    6th, 15, 19, 34, 35, 41, 48, 52, 61, 64, 65
    7th, 15, 25, 35, 52, 61, 63, 64
    8th, 15, 19, 35
    9th, 13, 15, 19,
    10th, 15
    11th, 15, 35, 41, 48, 53
    12th, 15, 48, 53
    13th, 41, 53
    14th, 35
    15th, 35
    16th, 53
Kentucky Mounted Infantry:
    4th, 48, 52, 61, 65
    11th, 35
    30th, 53
    37th, 53
    39th, 53, 64
    40th, 53
    45th, 41, 53
    52nd, 53
    54th, 53
Kentucky, State of, 13, 14, 72, 79, 90
Kentucky (3rd Union), 12
Keough, Capt. Myles, 104
Kettle Hill, 116, 124, 126
Kilpatrick, Judson, 56, 66
Kiowa Indians, 50
Kirk, W. Stokes, 81
Kirkland Bros., 80, 81
Kittredge, Ben & Co., 13, 17, 34
Klamath, Fort (Oregon), 96
Kongberg Armory, 119
Krag, Capt. Ole, 119
Krag carbines, 118-124, 126-130
Krag-Jorgensen rifles, 115, 119
Krag rifles, 120, 121, 123, 126, 130

L

Lancaster, Fort (Texas), 89
Laramie, Fort (Montana), 69, 88
Lawton, Capt. Henry, 109, 124
Leavenworth Arsenal, 70, 75, 86-88, 92, 94, 97, 110
Leavenworth, Fort (Kansas), 127
Lee, Fitzhugh, 30, 46
Lee, Gen. Robert E., 60, 70, 75
Lee rifles, 107
Lieb, Edward H., 13
Lincoln, Abraham, Fort (North Dakota), 106, 110
Lindner carbines, 23-25, 35, 39, 81, 86
Lindsay rifle musket, 80
Little Bighorn, Battle of, 102, 103
Long, Col. Eli, 34, 61, 62
Louisiana Cavalry:
    1st, 19, 35, 53, 63

    2nd, 53
Lowell, Charles Jr., 49

M

Mackenzie, Ranald, 58, 59, 90
Maine Cavalry:
    1st, 19, 30, 32, 36, 47, 53, 65
    2nd, 64, 65
Manassas, Battle of, 12
Maryland Cavalry:
    1st, 19, 30, 32, 36, 53, 63
    2nd, 36, 63
    3rd, 63, 64
Massachusetts Arms Co., 27
Massachusetts Cavalry:
    1st, 19, 20, 27-29, 32, 36, 47, 53, 63
    2nd, 19, 36, 41, 53, 63, 64
    3rd, 36, 53, 63, 72
    4th, 53, 63, 65
Massachusetts Mounted Infantry:
    31st, 53
    40th, 53
Maus, Lt. Marion, 108
Mauser rifles, 124
Maynard carbines, 11-13, 15, 17, 35, 39, 51, 57, 58, 61, 67-69, 75, 77, 78, 80, 81, 87, 89, 90, 92, 93, 95, 105
Maynard cartridges, 81, 94
McCloud, Joe, 22
McConnell, Col. John, 65
McCook, Edward, 61
McCulloch, Col. Robert, 51
McIntosh, Col. John, 32, 47, 49
McKinley, Pres. William, 120
McLaughlin Ohio Battalion, 21
McLaughton Squadron, 64
Mead, Fort (South Dakota), 127
Meriden Mfg. Co., 72
Merrill carbines, 10, 14, 17, 19, 21, 24, 25-27, 30, 32-37, 39, 41, 42, 48, 51, 52-54, 57, 58, 60, 67, 72, 75, 79, 80, 85, 87, 93, 133, 136
Merrill cartridges, 94
Merritt, Gen. Wesley, 32, 46, 49, 60
Merwin and Bray, 13
Mexico, 77
Meyer, Fort (Virginia), 127
Michaelis, Capt. Otho E., 105, 106
Michigan Cavalry:
    1st, 13, 19, 32, 36, 46, 50, 53, 65
    2nd, 13, 19, 29, 36, 48, 53, 61, 65
    3rd, 19, 36, 42, 53, 65
    4th, 34, 48, 53, 61, 65
    5th, 29, 32, 33, 39, 45, 46, 53, 65
    6th, 32, 33, 36, 45, 46, 50, 53, 65
    7th, 32, 36, 46, 50, 53, 65
    8th, 34, 48, 53
    9th, 34, 48, 53, 64, 66
    10th, 53
    11th, 53
Michigan Mounted Infantry:
    1st, 45
Miles, Brig. Gen. Nelson, 107-109, 112, 120
Miller, Col. Abraham O., 62
Minnesota Cavalry:
    1st, 19, 36
    2nd, 53, 63
Minnesota Mounted Infantry:
    8th, 53
Minnesota Independent Battalion, 36
Minnesota Independent Cavalry, 53
Minty, Col. Robert H.G., 34
Mississippi Cavalry:
    1st, 36, 53, 63
    18th, 51
    19th, 51

Mississippi rifles, 51
Missouri Cavalry:
    1st, 36, 53, 63
    2nd, 19, 21, 36, 52
    2nd (Merrill's Horse), 25, 41, 42
    3rd, 21, 36, 53
    4th, 21, 36, 44, 53
    5th, 19, 64
    6th, 19, 36, 53
    7th, 19, 36, 53
    7th Volunteer, 24
    8th, 36, 45, 53
    8th Volunteer, 45
    10th, 36, 53, 61, 62, 65
    10th Volunteer, 25
    11th, 26, 36, 53
    12th, 19, 53, 61, 68
    12th Volunteer, 42
    13th, 19, 53
    14th, 53, 64
    15th, 64
    16th, 63
Missouri Mounted Infantry:
    32nd, 53
Missouri State Militia:
    1st, 36, 53, 63, 64
    2nd, 53
    3rd, 36, 53, 63
    4th, 36, 53
    5th, 36
    6th, 24, 36, 53, 64
    7th, 53
    8th, 53, 63
    9th, 53, 65
Modac War, 82, 96, 97
Monroe, Fort (Virginia), 86, 87
Morgan, John H., 33
Morgan's Raid, 33
Mosby's Rangers, 25, 26
Mount Vernon Arsenal, 86, 87
Mounted rifles, 11
Murray, Eli H., 12
Musketoon cartridges, 94
Musketoon, Model 1847 Cavalry, 16, 17
Musketoons, 35, 57, 58, 87
Muskets .69 cal., 51

N

Nebraska, 1st Cavalry, 53
Nevada Cavalry:
    1st, 36, 54
    1st Battalion, 63
Nevada Volunteer Cavalry, 127
New Hampshire, State of, 70
New Hampshire, 1st Cavalry, 53, 63
New Jersey Cavalry:
    1st, 21, 32, 36, 47, 53, 69
    2nd, 29, 40, 53, 63, 65
    3rd, 47, 49, 53
New Mexico, 1st Cavalry, 21, 36, 50, 53
New Mexico, State of, 12
New York Agency, 77, 79, 80, 86, 87
New York Arsenal, 23, 70, 72, 79, 86-88, 93
New York Cavalry:
    1st, 21, 30, 36, 65
    1st Lincoln, 54, 64
    1st Mounted Rifle, 64, 65
    1st Veteran, 54, 64
    2nd, 21, 30, 32, 36, 47, 49, 54, 64, 65
    2nd Mounted Rifle, 64
    2nd Veteran, 54, 63
    3rd, 21, 36, 54, 65
    4th, 21, 32, 36, 42, 46, 63
    5th, 21, 32, 36, 47, 49, 54, 64
    6th, 21, 30, 32, 36, 46, 63
    7th, 21, 36, 54
    8th, 21, 30, 32, 36, 47, 63, 64
    9th, 21, 26, 27, 30, 32, 36, 39, 46, 54, 64
    10th, 21, 27, 30, 32, 36, 46, 54, 64
    11th, 21, 36, 54, 64
    12th, 36, 42, 54
    13th, 25, 26, 36, 54, 67
    13th Heavy Artillery, 67, 68
    14th, 36, 54, 64
    15th, 54, 63
    16th, 25, 36, 54, 63
    18th, 36, 54, 64
    19th, 24, 25, 33, 36, 46, 54, 64
    20th, 54
    21st, 54, 64
    22nd, 54, 63, 64
    23rd, 36, 54
    24th, 54, 63, 64
    25th, 50, 54, 64, 65
New York, State of, 12
Nez Perce uprising, 110
Nobles, Capt. George, 65
Nolan Brothers, 81
Norris, Samuel, 70

O

Officer's Model Rifles, 112
Ohio, State of, 12
Ohio Cavalry:
    1st, 15, 21, 34, 36, 48, 54, 61, 65
    1st Squadron, 15
    2nd, 15, 21, 34, 36, 47, 49, 54, 65
    2nd Independent Battalion, 15
    3rd, 15, 21, 34, 36, 48, 54, 61, 63, 65
    4th, 15, 21, 34, 36, 48, 54, 61, 62, 63, 65
    4th Battalion, 36
    5th, 12, 15, 17, 21, 36, 48, 54
    5th Battalion, 36
    6th, 15, 21, 24, 32, 36, 47, 54, 63, 64
    7th, 15, 36, 48, 54, 61, 64, 65
    8th, 54, 64, 65
    9th, 36, 54, 65, 66
    10th, 26, 36, 48, 54, 63, 65
    11th, 29, 36, 54, 65
    12th, 54, 65
    13th, 63
Ohio McLaughlin Squadron, 21, 48, 54, 132, 134
Omaha Depot, 97, 110
Opequon Creek, 49
Oregon, 1st Cavalry, 54

P

Pace, Lt. Col. Thomas N., 24,
Paiute Indians, 89
Palmer carbines, 69, 70, 78, 81, 86
Parks (gunboat steamer), 68
Pawnee Scouts, 92
Peking (China), 130
Pennington, Alexander, 59
Pennsylvania Cavalry:
    Greenfield Co., 21
    1st, 12, 21, 30, 32, 36, 47, 63
    2nd, 21, 36, 47, 63
    3rd, 12, 32, 36, 54, 55, 63, 64
    4th, 21, 32, 36, 47, 65
    5th, 21, 36, 54, 63, 65
    6th, 12, 30, 32, 34, 36, 46, 63
    7th, 11, 34, 36, 48, 54, 61, 62, 65
    8th, 21, 32, 36, 47, 54, 63
    9th, 13, 26, 36, 44, 54, 64-66

Index

11th, 21, 36, 54, 65
12th, 36, 42, 54, 64, 70
13th, 21, 36, 47, 54, 63, 64
14th, 21, 36, 54, 64, 65, 67, 70
15th, 36, 54, 63
16th, 21, 28, 32, 33, 36, 40, 47, 54, 65
17th, 30, 32, 36, 46, 54, 63, 65
18th, 32, 36, 47, 49, 54, 65
19th, 54, 63
20th, 36, 42, 54, 64
21st, 36, 54, 63
22nd, 36, 54
Percussion pistols, 51
Perry carbines, 79
Philadelphia City Troops, 120
Philippine Insurrection, 127
Philippines, 126, 129, 130
Pickett, George, 58, 60
Pine Ridge, 112
Pistol carbine cartridges, 94
Pistol carbines, 17, 35, 39
Pond, C.W., 76
Potomac Home Brigade, 36
Potomac Home Guard, 21, 53
Presidio (California), 127
Puerto Rico, 120, 126
Purnell Legion, 32, 36

R

Read, Wm. & Sons, 76
Red River Campaign, 102
Remington carbines, 70, 71, 73, 75, 78, 79, 87, 97
Remington cartridges, 94
Remington, E. & Sons, 70, 77, 78
Remington M1870 experimental carbines, 98, 99
Remington M1863 rifle, 80
Remington revolvers, 14, 88
Remington .36s, 30
Reno (gunboat steamer), 68
Reno, Major Marcus, 104, 105
Revolvers .36 cal., 51
Revolvers .44 cal., 51
Rhode Island Cavalry:
    1st, 10, 12, 21, 36, 49, 54, 63, 65
    3rd, 64
Rhode Island, State of, 12
Rifle muskets, 46, 61
Rifled Musketoons, 39
Rifled Pistol carbines, 75
Rifles (1841), 11
Riley, Fort (Kansas), 68
Ringgold Battalion, 21
Ripley, Gen., 12, 14, 26
Rolls, Capt. W.H., 29
Roosevelt, Lt. Col. Theodore, 116, 120, 121, 124
Ross, Brig. Gen. Lawrence, 50
Rough Riders, 116, 119, 120, 121, 123, 124, 126
Royall, William B., 13

S

Sabres, light cavalry, 88
Sackett, Col. William, 26, 39
Sam Houston, Fort (Texas), 127
St. Louis Arsenal, 59, 68, 70, 72, 77, 78, 86, 87, 92, 93
San Antonio Arsenal, 77, 92, 97, 101, 106, 108, 110
San Juan Hill, 124-126
Sanders, Samuel H. Co., 28
Santiago, 126
Savacool, Capt. Edwin F., 60
Savage pistols, 11
Savage Revolving Firearms Co., 70
Sayler's Creek, 60

Schoonmaker, Brig. James, 49
Schroeder carbines, 11, 15
Schyuler, Hartley and Graham, 12, 75, 76-78, 80
Sears Roebuck Co., 81
Selma Raid, 60, 62, 66
Shafter, Maj. Gen. William, 122, 124, 138
Sharps carbines, 11-14, 17, 18, 22, 28-35, 39-42, 44-51, 57-63, 65-68, 72, 74, 75, 77, 80-82, 84, 87, 89, 91, 93-95, 97-99, 102, 105, 120
Sharps cartridges, 81, 94
Sharps & Hankins carbines, 17, 26, 27, 30, 33, 35, 39, 46, 87, 93, 99, 105
Sharps & Hankins cartridges, 94
Sharps Model 1853 Slantbreech, 12
Sharps Rifle Mfg. Co., 12, 93
Sheridan, Gen. Phil, 29, 45, 46, 49, 50, 58-60, 102, 120
Sherman, Gen. James Harrison, 48, 51, 62, 88
Sherman's Cavalry, 56, 60
Sibley's Dakota Expedition, 22
Sill, Fort (Oklahoma), 84, 102
Sioux City Iowa Cavalry, 21
Sioux Indians, 92, 104, 112
Skeleton Canyon, 109
Skull Cave, Battle of, 97
Smith, Charles, 59
Smith carbines, 17, 20, 27-30, 33-35, 39, 44, 46-49, 51, 57, 58, 61, 62, 66, 67, 70, 75, 79-81, 87, 88, 93
Smith cartridges, 94
Smith & Wesson revolvers, 13, 14
Smooth-bore Musketoon, 39
Snelling, Fort (Minnesota), 110
Spanish-American War, 119, 120, 122, 124, 126
Spear, Samuel, 59
Spencer carbines, 39, 40, 46-51, 57-69, 72, 74, 76, 77, 79, 80-82, 87-90, 92, 93, 95, 99, 105, 120
Spencer cartridges, 24, 60, 65, 68, 69, 72, 77, 78, 81, 87, 88, 94
Spencer rifles, 29, 32-34, 46-48, 61, 62, 66, 78, 94
Spering, Nathan, 81
Springfield Arsenal, 79, 86, 87, 92-94, 97, 101, 102, 106, 108, 110, 119, 120, 127, 130
Springfield carbines, 16, 95, 97-99, 101, 102, 105, 106, 108-115, 120, 126
Springfield rifle muskets, 34, 46, 92, 111
Springfield rifles, 105, 107, 112, 123, 126, 131
Springfield shotguns, 112
Springfield (Trapdoor) carbines/rifles, 99, 100, 102, 114, 115, 119, 120, 122, 127
Stabler cut-off, 76, 77, 88, 90
Stagg, Peter, 59
Starr Arms Co., 73
Starr carbines, 35, 39, 42, 44, 46, 47, 49, 57-61, 67, 68, 70, 72, 73, 75-77, 79-81, 87, 88, 93, 105
Starr cartridges, 81, 94
Starr pistols, 11
Starr revolvers, 14
Starr, W.S., 77, 78
Stevens, Samuel, 60
Stoneman, George, 11, 12
Stuart, Jeb, 13, 29, 30, 31, 45
Symmes carbines, 11, 16, 77
Syms, Wm. & Bros., 75, 76

T

Tampa Ordnance Depot, 122, 123
Teal, John W., 13
Tennessee Cavalry:
    1st, 21, 37, 43, 48, 54, 63, 64
    2nd, 21, 37, 55, 64, 65
    3rd, 26, 37, 41, 55
    4th, 36, 55
    5th, 26, 37, 55
    6th, 37, 55, 63
    7th, 25, 37, 55, 64

8th, 55
9th, 55
10th, 37, 55, 63, 68
12th, 55, 61, 63, 65, 67, 68
13th, 55
Tennessee Mounted Infantry:
   1st, 55
   2nd, 55
   4th, 55
Terry, Brig. Gen. Alfred, 99
Texas Cavalry:
   1st, 21, 37, 55, 63, 64
   3rd, 51
   6th, 51
   9th, 51
   27th, 51
Thompson, Capt. John T., 122
Tientsin (China), 131
Todd's Tavern, 46
Tonto Basin War, 89, 96, 97
Torbert, Brig. Gen. Alfred T.A., 46, 49, 50
Town, Charles H., 13
Townsend, C.H., 76
Trapdoor Springfield carbines/rifles, 99, 100, 102, 114, 115, 119, 120, 122, 127
Trevilian Station, 45, 46
Triplett & Scott carbines, 68, 72, 73, 79, 80

U

Union carbines, 44, 51
Union, Fort, 85, 92
Union Pacific Railroad, 92, 93
U.S. Cavalry:
   1st, 11, 19, 30, 32, 35, 46, 52, 92, 95-98, 101, 102, 106, 107, 112, 120, 123, 124, 126, 127, 129
   1st at Leavenworth, Kansas, 12
   1st Colored, 52
   1st Regiment Volunteer Cavalry (Rough Riders), 119-121, 123, 124, 126
   2nd, 11, 19, 30, 32, 35, 46, 63, 72, 88, 89, 92, 94, 95, 98, 99, 101, 102, 108, 112, 123, 126
   2nd Colored, 26, 52
   3rd, 11, 19, 35, 51, 63, 92, 95, 98, 100-102, 120, 123, 126, 127
   3rd Colored, 52, 63, 64
   4th, 11, 19, 34, 35, 40, 48, 51, 62, 90, 92, 95, 98, 101, 102, 106-109, 111, 119, 127, 128
   4th Colored, 52, 63, 64
   4th Regular, 74
   5th, 11, 19, 30, 32, 35, 46, 63, 92, 95, 97, 98, 101, 102, 106, 107, 112, 126, 127
   5th Colored, 63
   5th at Gibson's Mill, 13
   6th, 11, 19, 30, 32, 35, 47, 63, 74, 92, 95, 98, 101, 102, 107, 109, 112, 120, 123, 126, 127, 130, 131
   6th Regular, 74
   7th, 88, 92, 95, 98, 99, 101, 102, 110-112
   8th, 88, 90, 92, 95, 98, 101, 102, 106-110, 112
   9th, 92, 95, 98, 99, 101, 102, 112, 120, 123, 126, 127
   10th, 84, 88, 90, 92, 95, 98, 101, 102, 106-108, 110, 112, 116, 123, 124, 126, 127
   11th, 127
   12th, 127
   13th, 127
   14th, 127
   15th, 127
U.S. Infantry:
   9th, 130
   14th, 89, 130
   15th, 130
   21st, 97
U.S. Artilleries:
   3rd, 130
   5th, 130

Upper Potomac Force, 63
Upton, Emory, 61
Ute Indians, 50

V

Valley Campaign, 49
Vancouver Arsenal, 69, 87, 92, 106, 110
Vermont, State of, 70
Vermont, 1st Cavalry, 12, 21, 32, 37, 47, 55, 63, 65
Viergertz, O.H. & Co., 75

W

Wainwright, Lt. Col., 69
Walcott, W.W., 76
Wallace, Lt. Col., 41
Ward-Burton M1870 experimental carbines, 98, 99
Warner carbines, 39, 57, 58, 61, 72, 73, 77, 78, 87
Warner cartridges, 94
Washington Arsenal, 86, 87, 92
Watertown, 86, 87
Watervliet, 86, 87, 92
Watkins, Col. Louis, 34
Webb, Lt. Col. Alexander, 28
Weld, W.F., 76
Wells, William, 59
Wesson carbines, 13, 16, 17, 34, 39, 57, 58, 68, 75, 77, 87
Wesson cartridges, 81, 94
West, Robert, 59
West Virginia Cavalry:
   1st, 21, 29, 37, 55, 63, 65
   2nd, 21, 37, 55, 65
   3rd, 30, 32, 37, 55, 63, 65
   5th, 55
   6th, 65
   7th, 23, 25, 37, 55, 65
West Virginia Mounted Infantry:
   2nd, 23
   3rd, 23
   8th, 23, 25
   10th, 23
Westherford, Col. J.W., 41
Wheeler, Maj. Gen. Joseph, 123, 124
Whipple, Fort (Arizona), 90
White Sulphur Springs, (West Virginia), 23
Whitney pistols, 11
Whitney revolvers, 14
Wichita Agency, 102
Wilson, Brig. Gen. James, 45, 47, 49, 60, 61, 62
Wilson Cavalry, 66
Wilson's Selma Raid, 66, 68, 72
Winchester, O.F., 77, 78
Winchester Repeating Arms Co., 106
Winchester rifles, 105, 108, 112
Winchester, Virginia, 49, 70
Windolph, Sgt. Charles, 105
Windsor Manufacturing Co., 71
Wisconsin Cavalry:
   1st, 26, 37, 48, 55, 61, 65, 72, 73
   2nd, 21, 37, 41, 55, 63, 64, 65
   3rd, 21, 37, 55
   4th, 37, 42, 55, 64
Wood, Col. Leonard, 120, 123
Woodstock Races, 50
Wounded Knee, 104, 108, 112

Y

Yavapais Indians, 97
Yellow Tavern, 45, 46